Is There Really a Difference?

The Chi-Square Test and the Null Hypothesis

Teacher's Guide

This material is based upon work supported by the National Science Foundation under award numbers ESI-9255262, ESI-0137805, and ESI-0627821. Any opinions, findings, and conclusions or recommendations expressed in this publication are those of the authors and do not necessarily reflect the views of the National Science Foundation.

Key Curriculum
1150 65th Street
Emeryville, California 94608
email: editorial@keypress.com
www.keycurriculum.com

First Edition Authors

Dan Fendel, Diane Resek, Lynne Alper, and Sherry Fraser

Contributors to the Second Edition

Sherry Fraser, Jean Klanica, Brian Lawler, Eric Robinson, Lew Romagnano, Rick Marks, Dan Brutlag, Alan Olds, Mike Bryant, Jeri P. Philbrick, Lori Green, Matt Bremer, Margaret DeArmond

Project Editors

Sharon Taylor

Consulting Editor

Mali Apple

Project Administrator

Juliana Tringali

Professional Reviewer

Rick Marks, Sonoma State University, CA

Calculator Materials Editor

Josephine Noah

Math Checker

Carrie Gongaware

Production Editor

Andrew Jones

Production Director

Christine Osborne

Executive Editor

Josephine Noah

Mathematics Product Manager

Timothy Pope

Publisher

Steven Rasmussen

Contents

Introduction

Is There Really a Difference? Unit Overview

Intent

The major focus of this unit is on determining whether a difference that shows up between samples from two populations implies, in a statistical sense, that there is a difference in the populations from which the samples come. Put another way, students use statistical techniques to determine the likelihood that an apparent difference may in actuality be nothing more than a normal fluctuation in sampling.

Mathematics

The unit explores two categories of problems:

- Problems that compare a single population to a theoretical model (the theoretical-model case)
- Problems that compare two distinct populations (the two-population case)

Students learn that statisticians often presume that a "neutral" hypothesis, called a **null hypothesis,** holds unless there is clear evidence to the contrary. In the context of the two categories of problems, the null hypothesis is that the single population *does* fit the model or that the two populations being studied *are* the same. Students learn that to evaluate the null hypothesis, they must examine whether the observed data could reasonably have occurred under that null hypothesis.

In the course of studying such questions, students will

- work with double-bar graphs to explore data
- form hypotheses and corresponding null hypotheses
- develop an intuitive sense for evaluating differences between sets of data
- learn ways of organizing and presenting data
- learn about designing and carrying out statistical studies

This unit builds on students' prior experience with statistical ideas in the Year 1 unit *The Pit and the Pendulum.* In that unit, students worked with the **normal distribution** and used the **standard deviation** statistic as their primary tool. In this unit, students use the **chi-square statistic,** or χ^2 statistic. In the main activities of the unit, students use the χ^2 statistic only in the case of one degree of freedom. Supplemental activities explore more general use of the statistic.

Although the unit makes intensive use of the χ^2 statistic, the real emphasis is on broader statistical ideas, such as the null hypothesis, **sampling fluctuation,** and **hypothesis testing.**

The main concepts and skills that students will encounter and practice during the course of this unit are summarized by category here.

Setting Up Statistical Investigations

- Distinguishing between data snooping and hypothesis testing
- Describing the characteristics of a good sample
- Making null hypotheses
- Using proportional reasoning to analyze the consequences of a null hypothesis
- Designing and conducting statistical experiments

Interpreting Data

- Making hypotheses about larger populations by analyzing sample data
- Constructing and drawing inferences from charts, tables, and graphs, including frequency bar graphs and double-bar graphs
- Determining whether to accept or reject a null hypothesis
- Understanding the consequences of rejecting a null hypothesis
- Interpreting statistical experiments and communicating the outcomes

The χ^2 Statistic

- Developing intuition about the meaning of the χ^2 statistic
- Using simulations to estimate the χ^2 distribution
- Interpreting the χ^2 distribution curve as a probability table
- Calculating and interpreting the χ^2 statistic in order to compare data from real-world situations to theoretical models
- Calculating and interpreting the χ^2 statistic in order to compare two populations
- Using the χ^2 statistic to make decisions
- Understanding some limitations in applying the χ^2 statistic

Related Concepts

- Working with conditional probabilities
- Using simulations to develop intuition and to obtain data about sampling fluctuation
- Developing intuition about when differences in samples indicate that the larger populations are likely to be different
- Understanding why neither numeric difference nor percentage difference is an adequate tool for measuring the "weirdness" of data

- Reviewing the normal distribution and standard deviation and their applications to decision making

Progression

The first few activities introduce double-bar graphs, data snooping, hypothesis making and testing, and sampling fluctuation. After seeing examples of the two categories of problems in the unit (comparing one population with a theoretical model and comparing two populations), students learn the importance of a null hypothesis and practice developing hypotheses that are appropriate for given situations. They also explore sampling fluctuation using loaded dice and develop an intuitive sense for the "weirdness" of data.

After reviewing standard deviation, students learn to use the χ^2 statistic and probability table, first in the context of the theoretical-model case and then in the two-population case. Their work in this unit culminates in *POW 9: A Difference Investigation,* which they will work on in pairs. They will propose a hypothesis about two populations that they think really differ in some respect, collect sample data about the two populations, and analyze the data using bar graphs, tables, and the χ^2 statistic.

Data, Data, Data

Coins and Dice

A Tool for Measuring Differences

Comparing Populations

POW Studies

Pacing Guides

50-Minute Pacing Guide (28 days)

Day	Activity	In-Class Time Estimate
1	*Data, Data, Data*	5
	Stick-up Graphs, Activity 1	30
	Introduce: *POW 7: A Timely Phone Tree*	10
	Homework: *Samples and Populations*	5
2	Discussion: *Samples and Populations*	10
	Stick-up Graphs, Activity 2	20
	Try This Case	20
	Homework: *Who Gets A's and Measles?*	0
3	Discussion: *Who Gets A's and Measles?*	15
	Try This Case (continued)	10
	Group work: *POW 7: A Timely Phone Tree*	20
	Homework: *Quality of Investigation*	5
4	Discussion: *Quality of Investigation*	10
	Coins and Dice	0
	Two Different Differences	40
	Homework: *Changing the Difference*	0
5	Discussion: *Changing the Difference*	10
	Stick-up Graphs, Activity 3	40
	Homework: *Questions Without Answers*	0

6	Stick-up Graphs, Activity 4	15
	Discussion: Questions Without Answers	10
	Loaded Dice	20
	Homework: Fair Dice	5
7	Presentations: POW 7: A Timely Phone Tree	15
	Discussion: Fair Dice	10
	Loaded or Not?	25
	Homework: The Dunking Principle	0
8	Stick-up Graphs, Activity 5	15
	Discussion: The Dunking Principle	10
	How Different Is Really Different?	25
	Homework: Whose Is the Unfairest Die?	0
9	Stick-up Graphs, Activity 6	15
	Discussion: Whose Is the Unfairest Die?	20
	Homework: Coin-Flip Graph	15
10	Discussion: Coin-Flip Graph	15
	A Tool for Measuring Differences	0
	Discussion: Normal Distribution and Standard Deviation	10
	Bacterial Culture	25
	Homework: Decisions with Deviation	0
11	Discussion: Decisions with Deviation	30
	Introduce: POW 8: Tying the Knots	15
	Homework: The Spoon or the Coin?	5
12	Discussion: The Spoon or the Coin?	10

	Measuring Weirdness	35
	Homework: *Drug Dragnet: Fair or Foul?*	5
13	Discussion: *Drug Dragnet: Fair or Foul?*	10
	How Does χ^2 Work?	40
	Homework: *The Same χ^2*	0
14	Discussion: *The Same χ^2*	10
	Measuring Weirdness with χ^2	40
	Homework: *χ^2 for Dice*	0
15	Discussion: *χ^2 for Dice*	10
	Does Age Matter?	30
	Homework: *Different Flips*	10
16	*Does Age Matter?* (continued)	10
	Discussion: *Different Flips*	10
	Graphing the Difference	30
	Homework: *Assigning Probabilities*	0
17	Discussion: *Assigning Probabilities*	10
	Random but Fair	30
	Reference: *A χ^2 Probability Table*	10
	Homework: *A Collection of Coins*	0
18	Discussion: *A Collection of Coins*	10
	Presentations: *POW 8: Tying the Knots*	15
	Introduce: *POW 9: A Difference Investigation,* Phase 1	20
	Homework: *Late in the Day*	0

19	POW 9: A Difference Investigation, Phase 2	5
	Discussion: *Late in the Day*	10
	Comparing Populations	0
	What Would You Expect?	35
	Homework: *Who's Absent?*	0
20	Discussion: *Who's Absent?*	10
	Big and Strong	30
	POW 9: A Difference Investigation, Phase 3	10
	Homework: *Delivering Results*	0
21	POW 9: A Difference Investigation, Phase 4	0
	Discussion: *Delivering Results*	20
	Paper or Plastic?	30
	Homework: *Is It Really Worth It?*	0
22	POW 9: A Difference Investigation, Phase 5	0
	Discussion: *Is It Really Worth It?*	10
	"Two Different Differences" Revisited	40
	Homework: *Reaction Time*	0
23	POW 9: A Difference Investigation, Phase 5	0
	Discussion: *Reaction Time*	10
	"Two Different Differences" Revisited (continued)	10
	Bad Research	30
	Homework: *On Tour with* χ^2	0
24	Discussion: *On Tour with* χ^2	10
	POW Studies	0

	POW 9: A Difference Investigation, Phase 6	40
	Homework: Wrapping It Up	0
25	Presentations: POW 9: A Difference Investigation, Phase 7	45
	Homework: Beginning Portfolio Selection	5
26	Discussion: Beginning Portfolio Selection	10
	Presentations: POW 9: A Difference Investigation (continued)	40
	Homework: "Is There Really a Difference?" Portfolio	0
27	In-Class Assessment	50
	Homework: Take-Home Assessment	0
28	Assessment Discussion	30
	Unit Reflection	20

90-minute Pacing Guide (21 days)

Day	Activity	In-Class Time Estimate
1	*Data, Data, Data*	5
	Stick-up Graphs, Activity 1	30
	Introduce *POW 7: A Timely Phone Tree*	15
	Try This Case	35
	Homework: *Samples and Populations*	5
2	Discussion: *Samples and Populations*	10
	Stick-up Graphs, Acitvity 2	25
	Who Gets A's and Measles?	50
	Homework: *Quality of Investigation*	5
3	Discussion: *Quality of Investigation*	10
	Coins and Dice	5
	Two Different Differences	35
	Group work: *POW 7: A Timely Phone Tree*	20
	Changing the Difference, Part I	20
	Homework: *Changing the Difference*, Part II	0
4	*Stick-up Graphs*, Activity 3	40
	Questions Without Answers	45
	Homework: *POW 7: A Timely Phone Tree*	5
5	Presentations: *POW 7: A Timely Phone Tree*	20
	Introduce: *POW 8: Tying the Knots*	15
	Stick-up Graphs, Activity 4	20

	Loaded Dice	30
	Homework: *Fair Dice*	5
6	Discussion: *Fair Dice*	10
	Stick-up Graphs, Activity 5	20
	Loaded or Not?	25
	The Dunking Principle	30
	Homework: *How Different Is Really Different?*	5
7	Discussion: *How Different Is Really Different?*	20
	Whose Is the Unfairest Die?	30
	Stick-up Graphs, Phase Activity 6	20
	Homework: *Coin-Flip Graph*	20
8	Discussion: *Coin-Flip Graph*	15
	A Tool for Measuring Differences	0
	Discussion: *Normal Distribution and Standard Deviation*	15
	Bacterial Culture	30
	Decisions with Deviation	25
	Homework: *The Spoon or the Coin?*	5
9	*Decisions with Deviation* (continued)	35
	Discussion: *The Spoon or the Coin?*	10
	Measuring Weirdness	45
	Homework: *POW 8: Tying the Knots*	0
10	Presentations: *POW 8: Tying the Knots*	15
	POW 9: A Difference Investigation, Phase 1	20
	How Does χ^2 Work?	45

	Homework: *Drug Dragnet: Fair or Foul?*	10
11	POW 9: *A Difference Investigation,* Phase 2	5
	Discussion: *Drug Dragnet: Fair or Foul?*	15
	The Same χ^2	40
	Measuring Weirdness with χ^2	30
	Homework: χ^2 *for Dice*	0
12	*Measuring Weirdness with* χ^2 (continued)	15
	Discussion: χ^2 *for Dice*	10
	Does Age Matter?	40
	Homework: *Different Flips* (begin in class)	25
13	POW 9: *A Difference Investigation,* Phase 3	15
	Discussion: *Different Flips*	10
	Graphing the Difference	30
	Random but Fair	35
	Homework: *Assigning Probabilities*	0
14	POW 9: *A Difference Investigation,* Phase 4	5
	Discussion: *Assigning Probabilities*	15
	Reference: *A* χ^2 *Probability Table*	15
	A Collection of Coins	45
	Homework: *Late in the Day*	10
15	POW 9: *A Difference Investigation,* Phase 5	0
	Discussion: *Late in the Day*	10
	Comparing Populations	0
	What Would You Expect?	40

	Who's Absent?	40
	Homework: *Big and Strong*	0
16	Discussion: *Big and Strong*	10
	Delivering Results	50
	Paper or Plastic?	30
	Homework: *Is It Really Worth It?*	0
17	*POW 9: A Difference Investigation,* Phase 5	0
	Discussion: *Is It Really Worth It?*	10
	"Two Different Differences" Revisited	50
	Bad Research	30
	Homework: *Reaction Time*	0
18	Discussion: *Reaction Time*	10
	On Tour with χ^2	35
	POW Studies	0
	POW 9: A Difference Investigation, Phase 6	45
	Homework: *Wrapping It Up*	0
19	Presentations: *POW 9: A Difference Investigation,* Phase 7	85
	Homework: *Beginning Portfolio Selection*	0
	Homework: *"Is There Really a Difference?" Portfolio*	5
20	Discussion: *Beginning Portfolio Selection*	15
	In-Class Assessment	50
	Homework: *Take-Home Assessment*	25
21	Assessment Discussion and Unit Reflection	40

Materials and Supplies

All IMP classrooms should have a set of standard supplies and equipment, and students are expected to have materials available for working at home on assignments and at school for classroom work. Lists of these standard supplies are included in the section "Materials and Supplies for the IMP Classroom" in *A Guide to IMP*. There is also a comprehensive list of materials for all units in Year 2.

Listed below are the supplies needed for this unit. General and activity-specific blackline masters are available for presentations on the overhead projector or for student worksheets. The masters are found in the *Is There Really a Difference* Unit Resources under Blackline Masters.

Is There Really a Difference?

- Colored stickers for stick-up graphs
- Poster paper or chart paper
- A bag large enough to store student-made dice
- Scissors (at least one for each pair of students)
- Cellophane tape and glue
- Dice (at least one per student)
- 1-foot lengths of string (six per pair of students)
- Coins (10 per pair of students)
- Spoons (between one to ten for each student)
- Chart graph paper
- Sentence strips (strips of paper roughly 3 feet by 3 inches, optional) for posting results

More About Supplies

- Sentence strips are useful in many IMP units. They are often used for posting solutions to problems, for posing problems, and for posting comments, strategies, and questions. These strips can be purchased at educational supply stores or simply made by cutting strips of construction paper, butcher paper, or chart paper.
- Graph paper is a standard supply for IMP classrooms. Blackline masters of 1-Centimeter Graph Paper, ¼-Inch Graph Paper, and 1-Inch Graph Paper are provided so you can make copies and transparencies. (You'll find links to these masters in "Materials and Supplies for Year 2" [link] of the Year 2 guide and in the Unit Resources of each unit.)

Assessing Progress

Is There Really a Difference? concludes with two formal unit assessments. In addition, there are many opportunities for more informal, ongoing assessments throughout the unit. For more information about assessment and grading, including general information about the end-of-unit assessments and how to use them, consult the *Year 2: A Guide to IMP* resource.

End-of-Unit Assessments

This unit concludes with in-class and take-home assessments. The in-class assessment is intentionally short so that time pressures will not affect student performance. Students may use graphing calculators and their notes from previous work when they take the assessments.

Ongoing Assessment

Assessment is a component in providing the best possible ongoing instructional program for students. Ongoing assessment includes the daily work of determining how well students understand key ideas and what level of achievement they have attained in acquiring key skills.

Students' written and oral work provides many opportunities for teachers to gather this information. Here are some recommendations of written assignments and oral presentations to monitor especially carefully that will offer insight into student progress.

- *Changing the Difference,* Part I: This work will give you information on students' sense of how probabilities behave with large samples.
- *Loaded or Not?:* This activity will tell you how well students can interpret experimental data.
- *Decisions with Deviation:* This assignment will provide information about students' understanding of how to use the normal distribution.
- *Measuring Weirdness with χ^2:* This activity will give you information about students' understanding of how to calculate and use the χ^2 statistic.
- *Late in the Day:* This assignment will give you feedback on how well students can set up and analyze a situation using the χ^2 statistic.
- *"Two Different Differences" Revisited:* This activity will give you information on students' abilities to do a complete analysis of a situation using the χ^2 statistic.

Supplemental Activities

Is There Really a Difference? contains a variety of activities at the end of the student pages that you can use to supplement the regular unit material. These activities fall roughly into two categories.

- **Reinforcements** increase students' understanding of and comfort with concepts, techniques, and methods that are discussed in class and are central to the unit.

- **Extensions** allow students to explore ideas beyond those presented in the unit, including generalizations and abstractions of ideas.

The supplemental activities are presented in the teacher's guide and the student book in the approximate sequence in which you might use them. Here are specific recommendations about how each activity might work within the unit. You may wish to use some of these activities, especially the later ones, after the unit is completed

***Two Calls Each* (extension)** This activity expands on the situation in *A Timely Phone Tree* and can be used either in place of that POW or as a follow-up activity.

***Incomplete Reports* (extension)** This open-ended activity about the use of surveys in the media can be assigned at any point in the unit, but students will probably do a better job after they have had a few days in class to think about surveys.

***Smokers and Emphysema* (extension)** This activity involves conditional probability and works well for following up *Drug Dragnet: Fair or Foul?*

***Explaining χ^2 Behavior* (extension)** This activity asks students to show how the χ^2 statistic reflects the effect of sample size. It can be assigned any time after *The Same χ^2* has been discussed.

***Completing the Table* (extension)** This activity, which explores the idea of linear interpolation in the context of the χ^2 probability table, can be used after *A χ^2 Probability Table* has been discussed.

***Bigger Tables* (extension)** This activity extends the ideas of expected numbers to larger tables than those students have been working with. Students are expected to answer Question 2 based on their intuition. They revisit the data in the supplemental activity *A 2 Is a 2 Is a 2, or Is It?* in which they are introduced to χ^2 probability tables for more than one degree of freedom.

***TV Time* (reinforcement)** The mathematics concepts in this activity are related to those used in *Paper or Plastic?* and *Is It Really Worth It?* The activity offers students more experience with finding expected numbers and using the χ^2 statistic in the two-population case.

Degrees of Freedom **(extension)** This activity can be used after *Paper or Plastic?* It introduces students to the concept of *degrees of freedom,* which is needed to extend the use of the χ^2 statistic to tables larger than 2-by-2.

A 2 Is a 2 Is a 2, or Is It? **(extension)** This activity continues work with the concept of degrees of freedom and can be assigned after students have completed the supplemental activity *Degrees of Freedom.*

Data, Data, Data

Intent

In these first few activities, students have some preliminary experiences with data. They create bar graphs, "data snoop," make hypotheses, and test their hypotheses through further data collection.

Mathematics

Statistical reasoning often involves generalizing from information about a **sample** to a conclusion about a larger group, or **population.** Part of that reasoning is to define the larger group carefully and determine whether a given sample can be thought of as a random sample from that group.

With persistence, one can find patterns in almost any set of data. Looking for such patterns is sometimes called *data snooping*. The careful researcher knows that patterns found in this way may be particular to the given data source. The challenge is to determine whether such patterns can be generalized.

Over the course of the unit, students gain experience with statistical methodology. In particular, they learn that although data snooping may suggest ideas, proper statistical work requires stating a hypothesis *before* gathering data to test it.

Progression

In *Data, Data, Data,* students use data about themselves to explore the issue of whether patterns in data can be generalized. In the teacher-led activities in *Stick-up Graphs,* they make several graphs to indicate how they personally fit into the categories related to a particular question. They make observations about these graphs, form hypotheses based on those observations, and test their hypotheses by examining related data from another class. The six activities in *Stick-up Graphs* will need to be spread out over approximately six days.

In *Samples and Populations,* students start with various "larger populations" and consider whether subpopulations created in particular ways would make good samples of those populations. Next students explore the broad issue of whether or when it is legitimate to make claims based on sampling information.

Stick-up Graphs (a series of teacher-led activities)

POW 7: A Timely Phone Tree

Samples and Populations

Try This Case

Who Gets A's and Measles?

Quality of Investigation

Stick-up Graphs

Intent

These six teacher-led activities offer students hands-on practice in gathering and interpreting statistical data.

Mathematics

To make "stick-up graphs," students place stickers on a graph to indicate how they personally fit into the categories related to a particular question (such as, "How many hours of TV do you watch on an average school night?"). They make observations about the resulting graph, form **hypotheses** based on those observations, and test their hypotheses by looking at the corresponding data from another class. Students learn that although their observations are valid for their own class, their hypotheses may not prove true, raising the issue of what larger population the class might represent as well as the issue of whether the class is a "good" sample for that larger population.

Progression

The work with stick-up graphs is divided into six activities that each need to take place on a separate day. Activity 3 is best undertaken immediately after the activities *Two Different Differences* and *Changing the Difference.*

Approximate Time

30 minutes for Activity 1

25 minutes for Activity 2

40 minutes for Activity 3

20 minutes for Activity 4

20 minutes for Activity 5

15 minutes for Activity 6

Classroom Organization

Whole-class discussion with group work

Materials

Poster paper (chart graph paper is ideal)

Colored stickers (sized to fit the graph paper)

Stick-up graphs from the paired class

Doing the Activities

Stick-up graphs are poster- or chart-paper graphs in which students indicate where they fit in some poll or survey by placing colored stickers in the appropriate columns, a process called "signing in." There are six stick-up graph activities in all, each of which should be presented on a different day. In addition, for each graph, each class should be paired with another class so that each class may make hypotheses based on their own data and then test those hypotheses based on a new set of data.

Activity 1: Data Snooping 1

The first stick-up graph asks the question, "How many hours of TV do you watch on an average school night (to the nearest half hour)?" Introduce the general idea of stick-up graphs and present today's example by displaying a blank graph something like this.

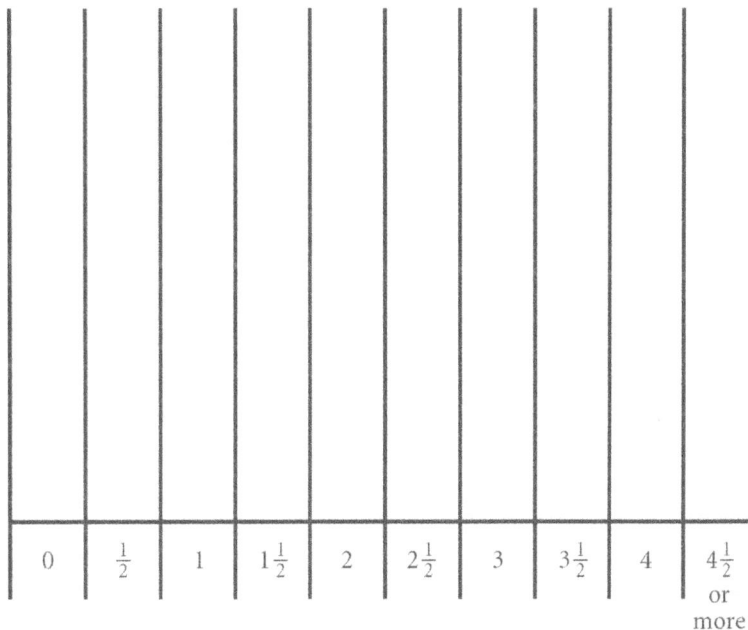

0	$\frac{1}{2}$	1	$1\frac{1}{2}$	2	$2\frac{1}{2}$	3	$3\frac{1}{2}$	4	$4\frac{1}{2}$ or more

Hours of TV watched on an average
school night, to the nearest half hour

Explain how students are to sign in, and have them file up to place their stickers on the graph. You will end up with a graph that looks something like this.

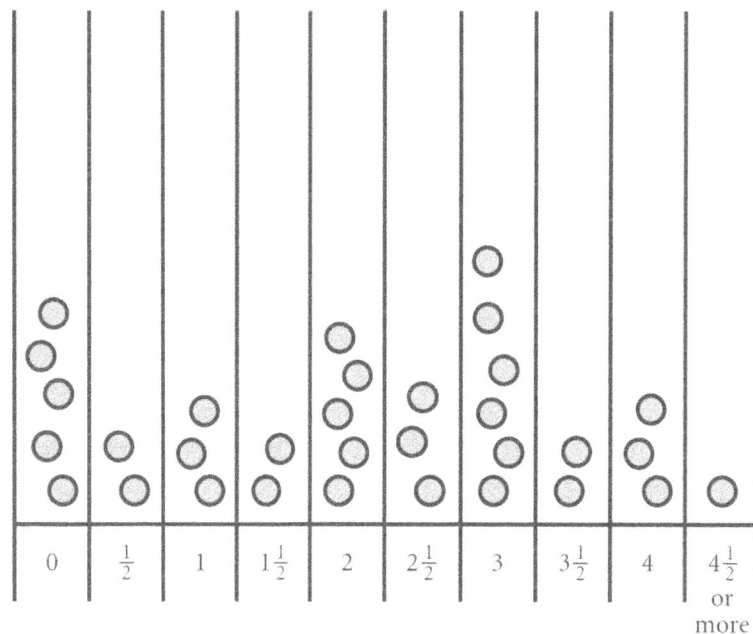

Hours of TV watched on an average
school night, to the nearest half hour

Finding Facts from the Graph

Have students suggest questions that might be answered from the graph or summarize information they can get from the graph. You might ask generic questions to get them thinking. What do you know based on the information in the graph? What larger population could this sample represent? What conclusions about the larger population can you draw from this sample? What conclusions might you be tempted to make but can't make reliably because the data aren't definitive?

Generalizing from the Sample

Raise the idea that a survey is usually intended to reveal something more than simply the opinions or responses of the people surveyed. To do this, you might ask, What is the goal of an election poll? Bring out that the pollster not only wants to know what the people polled think but also wants to use the results to predict what will happen on election day. What population or set of people is an election poll being used to represent? A variety of answers are possible, including "adults," "registered voters," and "likely voters." What's important is that students recognize that the polling sample is being used to represent some larger population.

With that idea in mind, return to the stick-up graph and ask, From what larger population can this class be considered a sample? Students should theorize that there are many such populations, such as

- all IMP students

- all high school students in the city, state, or country

- the population of the world

- the class itself (this is technically correct, although it is a special case)

The class represents some populations better than others, but it is a sample of all these populations nonetheless.

For which of these populations is the class a good sample? Emphasize that the goal of taking a sample from a population is to be able to say something about the *whole* population based on information about *part* of that population.

Data Snooping and Hypothesis Making

Tell students that the next step in their work with the stick-up graph is to propose some generalizations about a larger population, based on the sample data they have collected.

Begin by having the class decide on this larger population, which should include at least the students in both their own class and the paired class. For instance, it might be the set of all students at the school or all students at the school studying this unit.

Then have students suggest generalizations for this population. Bring out that any such generalizations will be partially guesswork and are thus called **hypotheses.** You may want to review this concept and students' previous experiences with hypotheses. For instance, they might think of a hypothesis as a prediction of how things are in the world in some respect.

Explain that students may come up with their hypotheses in a variety of ways, such as using their previous experiences and knowledge about their own and their friends' habits. However, emphasize the approach of basing hypotheses on the data in the graph. Students should look to their actual data for clues to possible generalizations. Tell them that this technique is called *data snooping* while the overall process of developing hypotheses is called *hypothesis making.*

Inform students that tomorrow they will find out how the paired class filled out the stick-up graph and will use those data to evaluate their hypotheses. Explain that this evaluation process is called **hypothesis testing.** Record students' hypotheses on chart paper, emphasizing the importance of stating them clearly.

Double-Bar Graphs

As a lead-in to the concept of double-bar graphs, ask the class whether the graph shows a difference between boys and girls with respect to amount of TV viewing. Students should respond that the graph doesn't give any information about how boys and girls compare.

Introduce the term *double-bar graph,* and describe the method of having two columns—one for boys and one for girls—for every time category on the current graph. Then have students create the framework for a double-bar graph for the TV-watching data.

B | G   B | G   B | G   B | G   B | G   B | G   B | G   B | G   B | G   B | G

0 $\frac{1}{2}$ 1 $1\frac{1}{2}$ 2 $2\frac{1}{2}$ 3 $3\frac{1}{2}$ 4 $4\frac{1}{2}$ or more

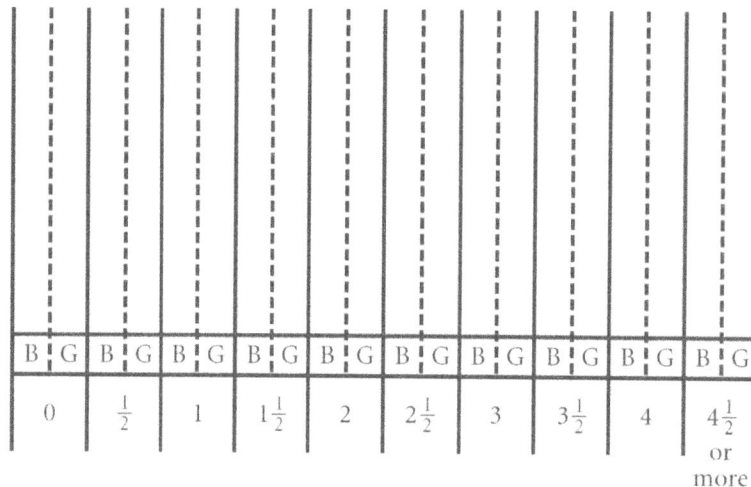

Hours of TV watched by boys and girls on an average
school night, to the nearest half hour

Tell students that in days to come they will choose a new topic suitable for a double-bar graph and sign in on that graph. They will go through the steps of data snooping, hypothesis making, and hypothesis testing for that graph as well.

Activity 2: Hypothesis Testing 1

In this activity, students will use the sample from the paired class to check their hypotheses from the previous discussion. Before looking at specific hypotheses, ask, **Why can't you check your hypotheses with your own data from the stick-up graph?** As needed, help them to articulate that their data on TV viewing led them to their hypotheses, so a *test* of their hypotheses is whether they hold true for a new sample.

Review the term *data snooping* for the process of examining data for possible hypotheses. As the word *snooping* often has a negative connotation, suggesting that something improper is being done, assure students that data snooping is a perfectly legitimate activity—they are "snooping" on data they have a right to review. Emphasize that data snooping is used to form hypotheses (*hypothesis making*), not to test them.

Also review the term *hypothesis testing* for the process of evaluating a hypothesis by finding out whether it holds true for a new set of data.

Display the stick-up graph from the paired class. As students examine whether a given hypothesis holds true for the new data, bring out that this still does not determine whether the hypothesis holds true for the larger population originally defined in the hypothesis (for example, all high school students). At this stage, students have no statistical basis for judging whether their hypothesis is true for that larger population, so their judgment will have to be based on intuition. The hypothesis might fit the new data but not hold true in general; conversely, it might not fit the new data yet still hold true in general. Encourage students to articulate clearly their analyses of whether their hypotheses are borne out by the new data.

Stages of Statistical Investigation

Ask students to try to summarize the stages involved in their work with the stick-up graphs so far. They might come up with something like this.

Data Collection—Collect sample data about a particular topic (with or without a particular idea of what might turn up).

Data Snooping—Examine the sample data, looking for interesting patterns.

Hypothesis Making—Form a hypothesis, either describing a population from which the sample was drawn or comparing two populations from which samples were drawn.

More Data Collection—Examine a new sample.

Hypothesis Testing—Examine whether the hypothesis fits the new data.

Evaluation—Decide whether you think the hypothesis is true.

Post this outline. You may want to use sentence strips for each stage to facilitate changes. This list will be refined in Activity 3, with a new step—formulating a null hypothesis—introduced between stages 3 and 4 and with stages 5 and 6 modified.

Activity 3: Data Snooping 2

Activity 3 is best undertaken immediately after *Two Different Differences* and *Changing the Difference,* as much of the discussion surrounding this stick-up graph will involve questions from those two activities. Because this discussion introduces the concept of **null hypothesis,** it is also necessary that this activity be completed before assigning *Questions Without Answers* or *The Dunking Principle.*

Give students a few minutes in their groups to share the questions from Part II of *Changing the Difference* that they would like to investigate and that would be suitable for a double-bar graph. Then bring the class together to discuss what types of questions lend themselves well to double-bar graph analyses. In preparation for introducing the concept of a null hypothesis, have students phrase their questions in a way that focuses on the comparison of two populations. Their questions should make it clear what populations are being compared and how they are being compared.

One approach to developing this phrasing is to give students this general model, which reflects the title of the unit:

Is there a difference between (population 1) and (population 2) with respect to (a particular characteristic)?

Then ask students how the question in "To Market, to Market" of Two Different Differences could be phrased in this way. They should come up with something like, "Is there a difference between women and men with respect to their preference for the new soft drink?" The characteristic here is "preference for the new soft drink," and a person either does or does not prefer the new drink.

Use this example to illustrate how to turn a question of this form into a double-bar graph. There are two possible outcomes for the characteristic, so there are two pairs of bars in the graph. The double-bar graph might look like this.

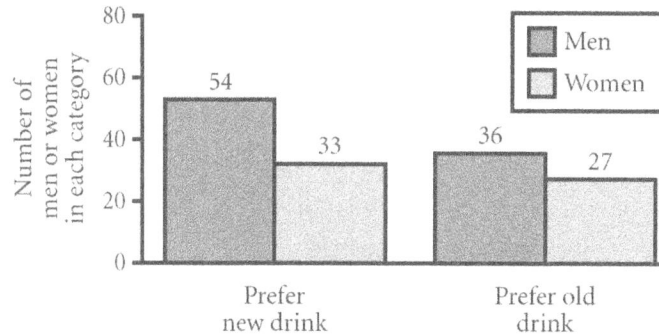

Rephrasing and Selecting Questions

Have the class decide on a topic for which data will be collected today and another topic for Activity 5. It will be nice if students choose topics for which they have no hypotheses prior to data collection. That will enable them to "snoop" their class data, hypothesize, and then check their hypotheses with a new sample.

The Null Hypothesis

Explain that in many situations, a researcher starts out with a neutral assumption that precedes the gathering of evidence. In statistics, this neutral assumption is known as the **null hypothesis.**

Explain that in the case of coin flips, the null hypothesis is usually that the coin is fair. Only if substantial evidence indicates that a coin is unfair do we discard this assumption. Tell students that discarding this assumption is called rejecting the null hypothesis.

Analogy with the Criminal Justice System

Tell students that the concept of a null hypothesis is similar to the concept of our criminal justice system, in which the fundamental presumption is that the accused is considered innocent under the law unless proven guilty. Ask, According to our criminal justice system, what is the null hypothesis about the defendant? Students should recognize that in the legal system, the null hypothesis is that the individual is not guilty. There must be evidence of guilt "beyond a reasonable doubt" before one rejects the null hypothesis of innocence and convicts an individual of a crime.

Why does our criminal system use this presumption of innocence? Help students to articulate that the criminal justice system was set up with this special presumption to decrease the chances of an innocent person being sent to jail.

To continue the analogy, ask, What is the prosecutor's hypothesis? Students should realize that the prosecutor's hypothesis is that the accused is guilty. Though this hypothesis is based on some evidence of guilt (otherwise there wouldn't be a trial), the burden of proof is on the prosecutor.

Although the concept of a null hypothesis does not lend itself easily to every statistical hypothesis, it will play a key role in this unit, and the two situations in Two Different Differences illustrate it well.

A Suspicious Coin

Ask, **What represents innocence in "A Suspicious Coin"?** The analogy with the legal system works well here, because Roberto is essentially accusing his brother of cheating. For the brother to be innocent means the coin is fair, so the null hypothesis is that Roberto's brother's coin is fair.

In line with the "Is there a difference?" theme, ask students if they can state this null hypothesis using the phrase "there is no difference." They should be able to come up with something like, "There is no difference between Roberto's brother's coin and a fair coin in terms of how often they come up heads and tails." You might point out that this null hypothesis is what Roberto would probably have assumed in the absence of any data.

What is Roberto's hypothesis? As "prosecutor," Roberto's hypothesis is that his brother is guilty, which means the coin is unfair. More specifically, his hypothesis is that the coin is biased in favor of heads.

Do you believe (or do you think Roberto believes) the null hypothesis? Some students may think this null hypothesis seems false. Ask them once again to articulate their reasoning, which may be something like, "A fair coin wouldn't give the results this coin is giving."

As was noted in the discussion of Changing the Difference, students might have different standards, and at this stage they are all working from their intuition about coin flipping. But they should agree that there is some kind of evidence they would consider strong enough to cause them to reject the null hypothesis.

Tell students that this analysis is the key. If the results that Roberto obtained would be "very unreasonable" for a fair coin, Roberto can conclude with some confidence that the coin is unfair. In other words, he can reject the null hypothesis. In this case, rejecting the null hypothesis means concluding the coin is "fixed."

Rejecting the Null Hypothesis

Presumably, if the coin is not fair, it is biased in favor of heads, although this conclusion comes from the data, not from anything in the null hypothesis. In general, rejecting the null hypothesis doesn't necessarily confirm any particular alternate hypothesis. But in rejecting the null hypothesis, one should suggest some alternate hypothesis under which the given results would have been more likely to occur. In this case, Roberto's results would be more likely if the coin were biased in favor of heads.

The courtroom analogy is helpful here. In many cases, jurors will vote guilty even though they think it is possible the defendant is innocent. They just think that their uncertainty about the defendant's guilt does not reach the standard of a "reasonable doubt." Statisticians do the same sort of thing. You may want to mention that the standards required for rejecting the null hypothesis may vary with the situation and may depend on what the consequences would be in each case.

If the evidence is insufficient to reject the null hypothesis, that fact does not constitute a demonstration that the null hypothesis is true. For instance, if we flip a coin 100 times and get 52 heads, we probably wouldn't reject the null hypothesis that the coin is fair. In fact, the coin might be slightly biased in favor of heads, but we don't have enough evidence to justify throwing out the assumption that the coin is fair.

Similarly, when a jury finds a defendant not guilty, it doesn't necessarily mean there is evidence of innocence or that there is no evidence of guilt; it simply means there is not sufficient evidence of guilt. This distinction will be discussed further as the unit progresses.

To Market, to Market

Now ask, **What is the null hypothesis for "To Market, to Market"?** Students will probably say that the null hypothesis is that the two populations—men and women—are the same in terms of soft drink preferences. Caution them to include the characteristic being studied. The null hypothesis is not, "Men and women are the same," but rather, "Men and women are the same in terms of their preferences for the two soft drinks."

Point out that this null hypothesis is typical in comparing two populations—the null hypothesis usually states that two populations are the same. If an experimenter hypothesizes that two populations are different, the burden of proof is on him or her.

Have students rephrase the null hypothesis using the "there is no difference" language. For example, "There is no difference between men and women with respect to their preferences for the new soft drink."

To clarify this null hypothesis, you might ask, **Does the null hypothesis say that people like the new soft drink and the old soft drink equally well?** Be sure students realize that the comparison under consideration is between men and women, not between the two drinks. In fact, the null hypothesis does not say anything about the likelihood that a given person will prefer the new soft drink. It says only that this likelihood is the same for men as for women.

What hypothesis is suggested by the sample data? Students should agree that data snooping suggests this hypothesis: "A man is more likely than a woman to prefer the new soft drink over the old one." They should also recognize that the proposed null hypothesis does not hold for the samples given in the situation. The question becomes, "Is it safe to conclude that the null hypothesis is also false for men and women in general?" Students should recognize that the preferences difference between men and women in the supermarket survey is not large.

As in the coin problem, the goal is to decide whether the information in the data set in the activity is strong enough to justify rejecting the null hypothesis. You may want to review the trial analogy here. The null hypothesis is equivalent to a verdict of not guilty. If you can't prove the accusation, the defendant goes free; if you can't justify rejecting the null hypothesis, you accept it for lack of any better conclusion.

Stages of Statistical Investigation Revised

Conclude today's discussion of the concept of a null hypothesis by amending the sequence of stages in a statistical investigation posted in Activity 2. The revised version might look like this.

Data Collection—Collect sample data about a particular topic (with or without a particular idea of what might turn up).

Data Snooping—Examine sample data, looking for interesting patterns.

Hypothesis Making—Form a hypothesis, either describing a population from which the sample was drawn or comparing two populations from which samples were drawn.

Null Hypothesis—Formulate an appropriate null hypothesis, usually stating that two populations are the same with respect to a particular characteristic. The specific characteristic will depend on the hypothesis made in Stage 3.

More Data Collection—Examine a new sample.

Hypothesis Testing—Find the probability that data like those in your new sample would have been obtained if the null hypothesis were true.

Evaluation—Based on the probability found in Stage 6, decide whether to reject the null hypothesis.

These are the key changes from the earlier version.

Stage 4 (Null Hypothesis) has been added.

Hypothesis testing and evaluation are now stated in terms of the null hypothesis and a specific probability.

Back to the Stick-up Graphs

Review the two topics chosen earlier for double-bar graphs. As a class, make two identical stick-up graphs for the first topic—one to be filled in by this class now and one for another class to fill in. If your current class is the second of two paired classes, they will also be completing the graph created by the first class.

When students are finished signing in, have them examine the results. You can use the "generic questions" suggested in Activity 1 to get them thinking about the significance of the graph.

Ask each group to come up with a hypothesis and a corresponding null hypothesis, recording them to check with data from the paired class. Formulating a null hypothesis should be straightforward, as it can take this form: "There is no difference between (population 1) and (population 2) with respect to (a particular characteristic)."

Activity 4: Hypothesis Testing 2

Have students complete the graph developed by the other class. (This will have been done along with Activity 3 in some classes.)

Testing Hypotheses

The class previously filled in a stick-up graph for a topic they chose and made a blank graph for another class to fill in. Each group made a hypothesis based on their own class data. Now display the filled-in graph from the other class. Have groups examine their hypotheses and discuss whether the sample data from the other class support those hypotheses. They should begin by considering whether the new data justify rejecting their null hypotheses.

Sampling Fluctuation

Keep in mind that the major focus of this unit is on answering the question, "If a sample is taken from each of two populations and the two samples differ, when does that difference imply (in a statistical sense) that there is a difference in the populations from which the samples come?"

Sometimes the two overall populations may really be the same with regard to a given characteristic, but normal sampling fluctuation will have created an apparent difference. For example, flipping two fair coins 50 times each may, and probably will, give different results. In this example, the populations are the sets of all potential flips for each of the two coins. The samples are the two sets of 50 flips.

In reference to today's stick-up graph discussion, raise the distinction between a difference due to sampling fluctuation from essentially identical populations and a sampling difference caused by a real difference in the populations. This idea will need to be revisited over and over again, as it takes some students a long time to understand it.

Evaluating the Results

Because each group had their own hypothesis and null hypothesis, the groups' conclusions will differ. Ask if any group rejected their null hypothesis—or at least seriously considered doing so. If so, have one such group of students explain their thinking.

The group will probably point to "large" differences in results between the samples from the two populations. They will argue, roughly, that these differences would be very unlikely if the two populations are the same with respect to the characteristic under study.

Some students may point out that no matter how large the difference in the samples, it is not a guarantee of a difference in the populations (unless the samples form very large portions of the populations). Nevertheless, they should realize that *the larger the difference* between the samples, the *less likely* it is that the populations from which they come are the same.

Tell students that later in the unit, they will learn a statistical test to calculate the probability of such an "accidental" difference.

Activity 5: Data Snooping 3

In Activity 3, students chose two topics for stick-up graphs. They worked with one of these topics in Activities 3 and 4, and now they will work with the other. As a class, make two stick-up graphs for the second topic—one to be filled in by this class now, and one for the paired class. If your current class is the second of two paired classes, they will also be completing the graph created by the first class.

When students have signed in, you may want to review the "Stages of Statistical Investigation" outline created in Activity 2 and modified in Activity 3. Then have each group come up with and record a hypothesis and a corresponding null hypothesis.

Activity 6: Hypothesis Testing 3

Have students complete the graph developed by the other class.

Then have groups examine the hypotheses they developed earlier and discuss whether the new data sample supports their hypotheses. They should begin by considering whether the new data set justifies rejecting their null hypotheses.

Students should be growing more comfortable with explaining their reasoning. Although they still lack the quantitative tools for a complete analysis, they should be able to articulate how unusual a given result would be were the null hypothesis true.

Key Questions

Activity 1

What questions can be answered from this stick-up graph? What information does it give you?

What is the goal of an election poll? Who are the polled voters supposed to represent?

From what larger population can this class be considered a sample? For which of these populations is the class a good sample?

How could you set up the stick-up graph to compare boys and girls?

Activity 2

Why can't you check your hypotheses with your own data from the stick-up graph?

Activity 3

What kinds of questions are appropriate for a double-bar graph?

According to our criminal justice system, what is the null hypothesis about the defendant?

Why does our criminal system use this presumption of innocence? What is the prosecutor's hypothesis?

What represents innocence in "A Suspicious Coin"?

What is Roberto's hypothesis? Do you believe (or do you think Roberto believes) the null hypothesis?

What is the null hypothesis for "To Market, to Market"? Does the null hypothesis say that people like the new soft drink and the old soft drink equally well?

What hypothesis is suggested by the sample data?

Activity 4

Did any group reject its null hypothesis? If so, why?

POW 7: A Timely Phone Tree

Intent

Students gain experience in approaching a challenging multistep problem, organizing their results, and communicating their thoughts clearly and convincingly.

Mathematics

This activity involves an information-transmission tree.

Progression

After students have had a few days to work on this POW, give them some time in class to share ideas in their groups. Schedule presentations about a week after the POW is assigned.

Approximate Time

5–10 minutes for introduction

1–3 hours for activity (at home)

20 minutes of group work midway through the activity

15–20 minutes for presentations and whole-class discussion

Classroom Organization

Individuals, with some group work and whole-class discussion

Doing the Activity

This POW asks students to find the number of people who could be contacted through a given phone tree in a specific amount of time. Some students will find this POW to be easy; others will not.

It will be useful to provide some time for students to share ideas in groups before the POW is due. Students should have done some preliminary work but may be having trouble dealing with the complexity of organizing the problem.

You might begin the group work by asking students how they are doing so far. Focus the discussion on ideas for keeping track of the information.

The day before the POW is due, give three students transparencies and pens to take home for preparing presentations.

Discussing and Debriefing the Activity

Have the three assigned students make their presentations.

Bring out that even though there is a correct answer (Leigh would be able to contact 1,048,575 friends through this phone tree), there are many ways to organize the data. Allow other students to explain their schemes, and ask whether anyone has a simpler method for finding the answer than those presented.

Supplemental Activity

Two Calls Each (extension) expands on the situation in *POW 7: A Timely Phone Tree* and can be used either in place of the POW or as a follow-up activity.

Samples and Populations

Intent

Students examine the idea of population sampling.

Mathematics

This activity focuses on examining whether a sample represents a population and constructing double-bar graphs.

Progression

Note: Be sure you have reviewed the teacher-led activities in *Stick-up Graphs.*

Students individually work on *Samples and Populations* and share ideas in a class discussion. Part I gives them several situations to consider. In each situation, students are asked to state whether a given sample is a good sample of the population and to give one example of an erroneous conclusion about the population that might be drawn from using that sample. These questions should make for lively discussion. Part II asks students to construct a double-bar graph.

Approximate Time

25 minutes for activity (at home or in class)

10 minutes for discussion

Classroom Organization

Individuals, then groups, followed by whole-class discussion

Doing the Activity

This activity requires little or no introduction.

Discussing and Debriefing the Activity

Have students share answers to Part I in their groups. Tell them that each group will share ideas on one question with the whole class. As you circulate, pay attention to students' double-bar graphs from Part II to decide how much discussion time to devote to them.

Start the discussion of Part I by asking members of one group whether they think Boy Scouts are a good sample of high school students and why. It's likely they will not think so. Two possible reasons are that boys who enjoy scouting might have different attitudes about music than do non-Scouts and that the sample includes no females.

Illustrate the meaning and use of the word **bias** in connection with these explanations. For instance, you might say the producer's study is biased toward male students.

For Question 1b, ask for an example of a conclusion the producer might reach based on a survey of Boy Scouts that might not be true about high school students in general.

For Part II, have presentations if you perceive a need to discuss the mechanics or ideas behind the making of double-bar graphs.

Try This Case

Intent

This activity gives students a context for the process of generalizing from a sample to a larger population. Students present arguments related to sampling fluctuation, based on proportional reasoning.

Mathematics

The situation in this activity is stated in terms of an *a:b:c:d* ratio that may be unfamiliar to students. Students must use proportional reasoning to generalize from a sample to a larger population. This requires consideration of **sampling fluctuation,** introducing the notion that there may be varying degrees of uncertainty as to whether sampling fluctuation is coincidental.

Progression

Students work on the activity in groups and share ideas in a class discussion.

Approximate Time

30–40 minutes

Classroom Organization

Groups, followed by whole-class discussion

Doing the Activity

You may want to review the ratio notation 4:3:2:1, perhaps starting with a simpler example, such as this one: *If the ratio of red to yellow marbles is 2:1, and there are 150 altogether of these two colors in a package, how many of each color would you expect to find?*

Give groups some time to work on this problem, and have someone present an analysis. You might then offer a more complex example. If the ratio of blue to yellow to green to red marbles is 8:4:2:1, and there are 150 altogether of these four colors in a package, how many of each color would you expect to find?

If students are having trouble with the arithmetic of Try This Case, you might discuss how many of each type of nut there would be if the 300 nuts were exactly in the ratio 4:3:2:1.

Discussing and Debriefing the Activity

Begin the discussion by polling the class to find how many groups would take each side of the case. Next, ask, What is Mr. Swenson's hypothesis? If students say something like, "He thinks he is getting ripped off," ask why he thinks that. They will probably respond with something like, "The nuts in his can weren't the numbers that were advertised."

Tell students that one way to say this is that Mr. Swenson thinks there is a difference between actual cans of Fresh Taste and cans that have a 4:3:2:1 ratio of nuts. At this point, you might have someone state how many nuts of each kind would be in each can if the contents fit the ratio exactly. You might bring out that if the total number of nuts was not a multiple of 10, it would be impossible to get whole-number values when the nuts are in the ratio 4:3:2:1.

Clearly, the can Mr. Swenson bought does not fit that ratio exactly. But the manufacturer would argue that its advertising refers to overall production, not to each individual can.

To counter this defense, Mr. Swenson would need to show that the overall "population" of nuts in all cans of Fresh Taste mixed nuts is not in the 4:3:2:1 ratio. In other words, he would argue that the overall Fresh Taste mixed nuts population differs from a 4:3:2:1 population. So far he has shown only that his can—which is a sample from this population—differs from a 4:3:2:1 population.

Have volunteers describe what their groups would accept as sufficient evidence to "convict" the manufacturer. Eventually, focus the discussion on exactly how much a sample of nuts needs to differ from a 4:3:2:1 ratio to conclude that the sample is not from a larger population with that ratio.

You might ask students to consider how they would react if Mr. Swenson had bought a can with, say, 150 peanuts, 120 walnuts, 20 almonds, and 10 cashews, or a distribution even further from the advertised ratio. Help them to articulate that some kinds of evidence would indicate that it is extremely unlikely Fresh Taste is really using a 4:3:2:1 ratio.

Ask students to envision a huge bin containing the four types of nuts in the precise 4:3:2:1 ratio. Have them imagine that each of them mixes the bin and chooses exactly 300 nuts at random. Then ask whether they would expect everyone to get samples exactly in the 4:3:2:1 ratio.

Ask students to explain what they think is going on here. Introduce the term **sampling fluctuation** to describe the kind of variation one would get in such an experiment.

Key Questions

What is Mr. Swenson's hypothesis?

Would you expect everyone to get samples exactly in the 4:3:2:1 ratio?

Who Gets A's and Measles?

Intent

This activity will help students realize the need to separate hypothesis testing from hypothesis making and will strengthen their understanding of the need for a two-stage data analysis—a data-snooping stage and a hypothesis-testing stage.

Mathematics

Data snooping may indicate associations or relationships that are actually nothing more than quirks of a particular sample. Observed associations may result simply from sampling fluctuations; they may be present in the particular sample but not occur in the larger population. Indeed, one can find strange associations in almost any sample. Only by testing the associations in a new sample can one have any statistically credible basis for believing they hold true in general.

Progression

Students work on their own to analyze the methodologies in two studies and to consider what can actually be concluded from each study. They share and justify their responses in a discussion that addresses the similarities and differences between hypothesis making and hypothesis testing.

Approximate Time

35 minutes for activity (at home or in class)

15 minutes for discussion

Classroom Organization

Individuals, followed by whole-class discussion

Doing the Activity

This activity requires little or no introduction.

Discussing and Debriefing the Activity

Begin by asking students for their ideas about the two situations. They will probably realize that Clara is foolish to think she can get an A by eating dessert, listening to rap music, and drinking juice, but they may not realize that the measles researcher may be making a similar error.

Students may assume the researcher has a physiological explanation for the link between weight and blood pressure and the occurrence of measles. However, no actual evidence indicates that this is the case.

This discussion concerns the distinction between data snooping and hypothesis testing. Even if one uses a new sample to test a hypothesis, how does one know whether the hypothesis can be generalized to the larger population? As students will learn, the key step is calculating the probability that a feature observed in a sample could have occurred even if the feature were not present in the overall population. Calculating such probabilities is what statistical tests are all about. The validity of such tests relies on the fact that the hypothesis is being tested with the sample, rather than that the sample is being used as the source of the hypothesis.

Ask, **Why might the medical researcher's data snooping have led to incorrect conclusions?** For example, the description does not tell how many measles patients were among the 500 patients studied or whether any patients in the sample were overweight or had high blood pressure but did not get measles. In addition, there is no mention of whether the researcher's sample is representative of the city's population. What if the researcher were snooping only through the records of cardiologists? That could explain a frequent occurrence of overweight patients with high blood pressure.

Although data snooping is not a reliable way to verify hypotheses, this doesn't mean the researcher's work is meaningless. Once students recognize the weaknesses in the researcher's methods, ask, **Is there any value in the medical researcher's study, despite its flaws?**

In many situations, people examine data so that hypotheses will occur to them. Medical research is one such area. It makes sense to look at lots of data to try to figure out what might be causing an epidemic, for example. There is nothing wrong with data snooping as long as it is used to *find* hypotheses rather than to *test* hypotheses. But then one needs to examine a different sample to determine whether one is looking at an anomaly of the sample or at a property of the larger population.

Key Questions

What are your overall impressions of these two situations?

Why might the medical researcher's data snooping have led to incorrect conclusions?

Is there any value in the medical researcher's study, despite its flaws?

Quality of Investigation

Intent

Data snooping may indicate associations or relationships that are actually nothing more than quirks of the particular sample. This activity reinforces the need to separate hypothesis testing from hypothesis making.

Mathematics

In any data set with enough potential variables, "accidental correlations" will appear in the data. One should not assume these correlations will hold true if the experiment is repeated. For a correlation to be useful, it must be something that can be expected in other samples or in a larger population.

Progression

Students work on the activity individually. Part I asks them to reflect on their positions on the controversy over conflicting studies regarding the dangers of smoking. Part II offers students a choice of activities, either locating and reviewing a published study or collecting their own data and giving an example of an incorrect generalization that could be made about some population from those data. The follow-up discussion reinforces the need to conduct multiple studies in order to have confidence in the conclusions.

Approximate Time

5 minutes for introduction

25 minutes for activity (at home or in class)

10 minutes for discussion

Classroom Organization

Individuals, followed by whole-class discussion

Doing the Activity

Clarify with students that they have a choice for Part II of the activity. You may want to have them brainstorm briefly to get ideas if they plan to do Option 2 of Part II. Any small sample from a large population will likely provide many opportunities for incorrect generalizations. You might suggest that students think about possible false generalizations before choosing their samples.

Discussing and Debriefing the Activity

You might let students share their thoughts about the issues raised in Part I within their groups. Bring out in the discussion that if enough studies are conducted on a hypothesis, occasionally one study will run counter to the truth. For example, if 100 people each roll the same unfair die repeatedly, some may get results that do not show an imbalance in the die.

Thus a single study is not necessarily definitive, no matter its outcome. Just as one study showing a correlation does not *prove* there is one, one study that does not show a correlation doesn't prove there *isn't* one. In the case of tobacco and cancer, many studies have been done, and the overwhelming majority show that smoking significantly increases the chances of getting cancer.

For Part II, you might have one or two students who chose Option 1 present a summary of the articles they read and suggest key missing information. You might then have one or two students who chose Option 2 describe their populations, their data, and the incorrect generalizations the data might suggest.

Supplemental Activity

Incomplete Reports (extension) is an open-ended activity about the use of surveys in the media.

Coins and Dice

Intent

These activities introduce the two situations that form the central problems for the unit. Students also continue their work with stick-up graphs and hypotheses.

Mathematics

The activities in *Coins and Dice* explore the concept of **sampling fluctuation.** A **sample** gives only a partial picture of a **population,** and each sample may be different. To know how much one can rely on the information from a sample, one needs to know how much variation to expect among samples.

These activities also introduce the concept of a **null hypothesis.** The basic presumption in much of statistics is that a new phenomenon is considered to be just like an old one unless there is convincing evidence otherwise. Thus a coin is presumed fair, and two populations are presumed similar, until data show the contrary. To make this presumption precise in a given situation, statisticians formulate a null hypothesis.

Progression

Two Different Differences introduces two situations showing data differences. In the first, the results from coin flips differ from the theoretical expectations for a fair coin. In the second, survey data in a sample of men differ from those for women. In each case, students are asked to decide, based on intuition, how significant these differences are. In *Changing the Difference,* they are asked what differences in these situations they would consider significant. Students then apply the concept of a null hypothesis to these two situations and conclude that they are fundamentally different: one involves comparison with theoretical probability; the other involves comparison of two populations.

In *Fair Dice,* students gather data to understand that even among fair dice, different samples will have different results, and to get a sense of how much fluctuation might occur among samples. They use this information to make judgments about possibly loaded dice in *Loaded or Not?*

Students then begin to explore the significance of sample size and recognize that a large sample with an "unusual" result is more significant than a small sample with a similar result. They investigate the competing criteria of sample size and "degree of unusualness" in determining how much significance to attach to an "unusual" data set.

Two Different Differences

Changing the Difference

Questions Without Answers

Loaded Dice

Fair Dice

Loaded or Not?

POW 8: Tying the Knots

The Dunking Principle

How Different Is Really Different?

Whose Is the Unfairest Die?

Coin-Flip Graph

Two Different Differences

Intent

Students do preliminary work with two situations that form the central problems for the unit.

Mathematics

This activity introduces the two basic types of problems that form the foundation of the unit: comparison of a single population with a theoretical model and comparison of two populations. These situations serve as prototypes for the theoretical-model case and the two-population case described in the unit overview. As students consider their confidence in hypotheses for the two situations, they recognize the need to quantify that confidence. The issue of **sampling fluctuation** arises again, and students obtain practice in identifying the populations and samples under consideration.

Progression

Students work in groups to examine two situations, one involving a coin that seems to be giving unexpected results and another involving a survey that seems to show a difference between men and women with respect to a particular criterion. In both situations, students are asked to state a hypothesis that supports a particular point of view, to state a hypothesis of their own, and then to rate how confident they are of their hypotheses. The follow-up discussion will lead students to articulate that the likelihood that a difference present in a sampling actually exists in the larger population increases with the size of that difference. It also reminds them of the need to be able to quantify the significance of the sampling difference.

Approximate Time

35 minutes

Classroom Organization

Groups, followed by whole-class discussion

Doing the Activity

No particular introduction is needed for this activity, so groups can get right to work.

Discussing and Debriefing the Activity

For each situation, have students from two or three groups read the group's report. At this stage, it is quite reasonable that students will have different opinions, and it is important to acknowledge the ambiguity of the situations. The upcoming activity *Changing the Difference* will help students to explore the gray areas involved. Keep a record of students' opinions at this stage (or have them keep their own records) so they can compare their intuitive ideas now with their opinions later, when they have statistical techniques to use.

A Suspicious Coin

Students should recognize that there is no absolute boundary between results that indicate the coin is unfair and those consistent with the assumption that the coin is fair. But they will presumably agree that as the results get more unbalanced, the less likely it is that a fair coin is being used. *Changing the Difference* will ask them to describe their "personal" boundary lines.

Tell students that one of the main goals of this unit is to characterize this decision-making process. Specifically, they will find a way to determine how likely it is that a fair coin would give various "unbalanced" results.

To Market, to Market

You may need to emphasize that the question being asked here is not "Which soft drink is preferred?" or "Who does most of the shopping?" but rather "Is there a difference between men and women in their reactions to the new soft drink?" It is also crucial for students to realize that the question is about men and women *in general* and not simply about the men and women who were surveyed.

Students should be able to articulate that in the sample data, the percentage of men preferring the new drink (60%) is higher than the percentage of women (55%). The crucial question is whether this difference in the sample data is enough to justify concluding that men and women in the general population differ in their attitudes toward the two drinks.

Some students may think the difference in the sample data reflects a genuine difference in the larger populations of "all men" and "all women." Others may think the difference is merely a normal sampling fluctuation. Bring out that a survey of men and women taken on a different day might have yielded a different result, just as one set of 10 flips of a given coin might produce 6 heads and 4 tails, while a different set of 10 flips of that same coin might produce 3 heads and 7 tails.

Populations and Samples

The key issue in most of the activities in this unit is whether differences that show up in the data for samples represent true differences in the populations from which the samples come. It will be helpful to review the distinction between the terms **sample** and **population** and to identify these in the situations from *Two Different Differences*. Ask, What populations are involved in these two situations? What are the samples from these populations?

The population and sample for "A Suspicious Coin" might be described this way.

The population is the set of all potential flips of the coin.

The sample is the set of 1000 coin flips Roberto made.

Students may be uncomfortable with the use of the word *population* to describe something other than a set of people. In "A Suspicious Coin," the population is made up of coin flips. You may need to explain that this broader usage is standard statistical terminology.

To clarify the issue, ask, **Which two populations are being compared?** Students should realize that Roberto is, in effect, comparing his brother's coin to a fair coin, so the second population is the set of all potential flips of a *fair* coin.

The populations and samples for "To Market, to Market" might be described this way.

The two populations are the set of all men and the set of all women.

The two samples are the set of men surveyed in the supermarket and the set of women surveyed in the supermarket.

Students may think of the survey as involving only one sample, consisting of all the people questioned. You may need to point out that to compare men's and women's reactions, students need to think of this set of people as containing two samples—one from the population of all men and one from the population of all women.

The samples from the two populations showed different percentages preferring the new drink. The key question is whether this difference in *samples* is true for the overall *populations.*

Could the observed differences between men and women be due to sampling fluctuations? Even if students do not ultimately agree on the answer to this question, they should recognize that the observed differences between men and women *could* be due to sampling fluctuations. They should begin to sense, however, that if the sampling results were more extreme, it would be very unlikely the differences were due to sampling fluctuations.

Inform students that in this unit they will learn concepts and techniques that will help them answer these questions and will return to these situations later with some mathematical tools for analyzing them.

Key Questions

What populations are involved in these two situations? What are the samples from these populations?

Which two populations are being compared?

Could the observed differences between men and women be due to sampling fluctuations?

Changing the Difference

Intent

Students develop their intuitive senses about the significance of sample differences.

Mathematics

The questions in Part I of this activity are posed to get students thinking about what issues are involved in deciding whether a sample difference indicates a difference between populations.

Progression

Part I asks students to decide on sample differences for several situations that would convince them that there really are differences in the populations. The follow-up discussion focuses on having students articulate that there is a relationship between the size of the sample difference and their confidence in their hypotheses.

The responses to Part II will be used in Activities 3 to 6 of the teacher-led exploration *Stick-up Graphs.*

Approximate Time

30 minutes for activity (at home or in class)

10 minutes for discussion

Classroom Organization

Individuals or groups, followed by whole-class discussion

Doing the Activity

Read the activity aloud. For Part II, you may want to discuss what "appropriate" means in this context.

Discussing and Debriefing the Activity

You might assign one of Questions 1 to 5 to each group for presentation to the class. As each question is presented, the class should debate the group's reasoning.

For Question 1, you may want to assure students that they are not expected to know how to calculate the probability of getting exactly 500 heads out of 1000 coin flips. (This probability is approximately .025, or about 1 in 40.)

For Question 2, you may need to do a little encouraging if the presenting group plays it safe by giving extreme examples. For part a, for example, you might encourage students to give as large a range of values as possible for which they would consider the coin fair (and not just something like "between 499 and 501"). The point that should come across for Questions 2 and 3 is that the further the coin results are from 50% heads, the more suspicious the coin is. Students should recognize that there is no precise transition point between "suspicious" and "not suspicious" results.

Similarly, for Question 4, urge students to do more than give extreme examples (such as data for part b in which nearly all the men prefer the new drink and nearly all the women prefer the old drink). They should realize that the greater the difference between the percentages for men and women in the samples, the more one would suspect that the overall populations—men and women—differ in their drink preferences.

For Question 5, students may point out a variety of similarities and differences. For the purposes of this unit, the key difference involves the nature of the two populations being compared in each situation. If necessary, review the discussion from *Two Different Differences* of what the populations are in each case. **What is compared in each situation?** Help students to focus on the distinction that in "A Suspicious Coin," one population is a theoretical model, while in "To Market, to Market," both populations are "real."

This distinction—comparing a population to a model versus comparing two populations—will surface later in the unit as students learn to use the chi-square test for statistical significance. It is not important that they completely grasp the distinction now.

Key Question

What is compared in each situation?

Questions Without Answers

Intent

Students propose hypotheses and null hypotheses for various situations.

Mathematics

This activity gives students further experience with formulating hypotheses and null hypotheses. The concept of a null hypothesis was introduced in Activity 3 of the teacher-led exploration *Stick-up Graphs.*

Progression

Students explore several situations that might suggest collecting data and are asked to formulate both a hypothesis appropriate to a particular perspective and a null hypothesis. The follow-up discussion focuses on the correct formation of a null hypothesis.

Approximate Time

35 minutes for activity (at home or in class)

10 minutes for discussion

Classroom Organization

Individuals, followed by whole-class discussion

Doing the Activity

This activity requires little or no introduction.

Discussing and Debriefing the Activity

Have students share their ideas about what the hypothesis and null hypothesis for each situation could be. There may be several reasonable choices for each question, especially for the hypotheses.

Without getting overly formal, encourage students to phrase their hypotheses and null hypotheses in terms of "there is a difference" and "there is no difference." The hypotheses should propose a specific difference.

For example, in Question 1, the executive's hypothesis might be, "There is a difference, with respect to sales of their releases, between artists who tour to promote their releases and artists who don't tour. Artists who tour have higher sales than artists who don't tour." The corresponding null hypothesis would be, "There is no difference, with respect to sales of their releases, between artists who tour to promote their releases and artists who don't tour."

Students may want to argue about the validity of some of their hypotheses and null hypotheses. Try to keep the focus of such discussion on the distinction between the hypothesis and null hypothesis.

This is another opportunity to reinforce the idea that one does not have to data-snoop to form a hypothesis. In each of these questions, with the possible exception of Question 2, the hypothesis seems to be based on prior experience or intuition. There is nothing wrong with this. However, researchers get into trouble when they data-snoop and then draw conclusions based on that snooping.

Loaded Dice

Intent

Students construct loaded dice in preparation for later data-gathering in the activity *Loaded or Not?*

Mathematics

This task provides a good opportunity to introduce (or review) the term *net,* a two-dimensional pattern that can be folded to form a three-dimensional object.

Progression

Students work in pairs to construct dice from construction paper. They "load" half of the dice by taping two paper clips to the inside of one face.

Approximate Time

25 minutes

Classroom Organization

Partners

Materials

Loaded Dice blackline master (copied on dark card stock; one per pair of students)

Large bag (for collecting and storing the dice)

Doing the Activity

Have each group of students split into two pairs. Each group will decide which pair will make a fair die and which will make a loaded die.

Collect the finished dice in a large bag and save them for use in *Loaded or Not?*

Discussing and Debriefing the Activity

No class discussion of this activity is needed.

Fair Dice

Intent

Students develop intuition about the significance of deviation from expected results by observing sampling fluctuation in experimental data from rolling a die.

Mathematics

In this activity, students will discover that equally likely events usually don't come up equally often in a given experiment and that they should expect some **sampling fluctuation.** They also consider how extreme such fluctuation must be to indicate that something more than coincidence is at work.

Progression

Students work on their own to roll a die 60 times, make a frequency bar graph of the data, and then repeat the process. The follow-up discussion brings out the variety of results in students' graphs and connects that variation with the concept of sampling fluctuation.

Approximate Time

5 minutes for introduction

30 minutes for activity (at home or in class)

10 minutes for discussion

Classroom Organization

Individuals, followed by whole-class discussion

Materials

Dice (one per student)

Doing the Activity

In preparation for Question 1, you may want to review the construction of a frequency bar graph, which students encountered in both *The Game of Pig* and *The Pit and the Pendulum* in Year 1. If you are assigning this activity as homework, inform students that they will need a die at home. (You might distribute them to students who don't think they have dice available at home.) Also, students could use their calculators to generate these random numbers; you may have to remind them about that procedure. You might suggest that the work will go faster if they have someone else tally as they roll. They can also roll several dice simultaneously, for a total of 60 rolls, to save time.

Emphasize that students' results are essential for use in the next activity, so it is especially important that everyone complete the activity and construct honest graphs.

Consider giving transparencies and pens to several students for preparing copies of their graphs. This will facilitate the subsequent discussion and provide more examples of graphs for comparison in the activity *Loaded or Not?*

Discussing and Debriefing the Activity

Have students share their frequency bar graphs for Question 1 in their groups. Each student should have two graphs, each representing 60 rolls of a die. Circulate around the room to check the graphs and look for possible areas of difficulty.

Then bring the class together and ask, Are any two of the graphs in a group identical? Did anyone get a "perfect" bar graph of ten 1s, ten 2s, and so on? It's very unlikely that either of these things happened. If some students prepared graphs on transparencies, have them show their results, and let the class comment on the variations.

Do any of your results make you doubt that your dice are fair? The key idea students should recognize here is that even with fair dice, they will get different frequency bar graphs. Review the use of the term **sampling fluctuation** to describe the variation in the graphs.

Ask additional questions to bring out the variation from sample to sample. For instance, Did anyone have a single number come up more than 15 times? More than 18 times? More than 20 times? Or, What is the highest "bar" in any of your frequency bar graphs? Students should recognize that even though they can expect the different numbers on a die to be rolled about equally often "in the long run," any particular experiment of 60 rolls will probably have unequal, and possibly very unequal, results.

What ideas do you have about how Lucky Lou in Question 2 might identify the loaded dice? Try to bring out the general idea that he might want to establish a guideline for how far off from "perfect" the results from a die would need to be before he labels it as loaded, which is about as much as one can say without more information. For example, a proper analysis of Lucky Lou's dilemma would require knowing the answers to such questions as these:

How many dice were there of each kind?

How were the loaded dice loaded? (For example, if the loaded dice always come up 1, it should be fairly easy to identify them quickly with little error.)

In the next activity, *Loaded or Not?,* students will simulate Lucky Lou's situation.

Key Questions

Are any two of the graphs in a group identical?

Did anyone get a "perfect" bar graph of ten 1s, ten 2s, and so on?

Do any of your results make you doubt that your dice are fair?

Did anyone have a single number come up more than 15 times? More than 18 times? More than 20 times?

What is the highest "bar" in any of your frequency bar graphs?

What ideas do you have about how Lucky Lou in Question 2 might identify the loaded dice?

Loaded or Not?

Intent

Students will roll a die and compare the frequency of the outcomes to that of a fair die to decide whether the die is loaded.

Mathematics

This activity demonstrates that equally likely events usually don't come up equally often in a given experiment and begins to develop students' intuition about the significance of deviation from expected results.

Progression

Students work in pairs to roll the dice constructed in *Loaded Dice* and try to determine which dice are loaded and which are fair. The subsequent discussion continues to build motivation for having a quantitative statistical tool to assist in making such determinations.

Approximate Time

25 minutes

Classroom Organization

Pairs, followed by whole-class discussion

Materials

Dice constructed in *Loaded Dice*

Frequency bar graphs constructed in Loaded Dice

Doing the Activity

Have students again break into pairs within their groups.

If several students have shown their frequency bar graphs from *Fair Dice,* the class will have a greater frame of reference as to what the graph for a fair die might look like.

Discussing and Debriefing the Activity

You might choose several students to read their responses. Then lead a discussion of how students are deciding whether their dice are loaded. Here are some possible questions to ask.

Is there any way to know for sure other than opening the die?

When are the results different enough to convince you that the die is loaded?

Would five paper clips (instead if two) make enough of a difference in the outcomes to show up in the data?

These questions are intended to elicit intuitive ideas, not precise answers. Remind students that they will soon be learning about a statistical tool that will allow them to do this kind of analysis more precisely. For now they should recognize that they need a general guideline about how different the results should be from those of a fair die before they can confidently conclude that the die is really different.

Let students open their dice at the end of class to find out if they were loaded.

Key Questions

Is there any way to know for sure other than opening the die?

When are the results different enough to convince you that the die is loaded?

Would five paper clips (instead if two) make enough of a difference in the outcomes to show up in the data?

POW 8: Tying the Knots

Intent

Students use a physical model to begin analysis of a combinatorial probability problem.

Mathematics

In this POW, students explore the possible outcomes of an experiment and the theoretical probabilities of the outcomes. As always, they are challenged to solve a difficult problem and to organize and communicate their process and solution clearly.

Progression

By way of introducing the POW, students perform the experiment in pairs and discuss their results. Whole-class presentations follow about a week later.

Approximate Time

15 minutes for introduction

1–3 hours for activity (at home)

15 minutes for presentations

Classroom Organization

Pairs, then individuals, followed by whole-class presentations

Materials

6 pieces of string, each about 1 foot long, per pair of students

Doing the Activity

Let students read the POW through the descriptions of the three stages. Then distribute six pieces of string to each pair of students and have pairs work through the three stages. When they are finished, bring the class together and have pairs describe their results.

Ask, What are the possible configurations once all the ends are tied? The goal is for students to discover that there are three possible outcomes:

Three small loops

One small loop and one medium loop

One large loop

It may be that no one gets three small loops. If this outcome (or one of the others) doesn't occur, ask students if they think there are any other possible outcomes. It's probably best for the development of the problem if students leave class at least aware of the three possibilities. Clarify that a given number of loops represents the same outcome whether or not the loops are interlocked.

Students have essentially just done Question 1 of the POW. Their task now is to find the probabilities associated with each possible outcome. Make sure they realize that they are looking for the theoretical probabilities, so an answer based solely on a simulation, or on experimental probabilities, will not suffice, although they might want to use simulations to test their analyses.

On the day before the POW is due, select three students to make presentations on the following day, and provide them with transparencies and markers.

Discussing and Debriefing the Activity

Have the selected students give their presentations. Other students may have arrived at their solutions in different ways.

Expect to get a variety of approaches. One possible approach to finding the probabilities is as follows. Number the strings from 1 to 6. Assume the upper knots have been tied and that 1 is tied to 2, 3 to 4, and 5 to 6. What matters now is how the lower knots are tied.

Key Question

What are the possible configurations once all the ends are tied?

The Dunking Principle

Intent
Students examine the effect of sample size on the interpretation of imbalanced results.

Mathematics
Students begin to explore the impact of sample size and realize that a large sample with an "unusual" result is more significant than a small sample with a similar result.

Progression
This activity presents a situation in which an event is supposed to be occurring 50% of the time, but appears to be happening more often. Students are asked to make judgments about how suspicious several sets of results are. The follow-up discussion explores the significance of sample size and the idea that a large sample with an unusual result is more significant than a small sample with a similar result; the larger the sample with an unusual result is, the more suspicious students should be that the population does not match the theoretical model.

Approximate Time
25 minutes for activity (at home or in class)

10 minutes for discussion

Classroom Organization
Individuals or groups, followed by whole-class discussion

Doing the Activity
Ask a student to read the activity aloud.

Discussing and Debriefing the Activity
You can probably go straight to Question 4, having two or three students share their decisions along with their reasoning. Students should realize that there is no right answer but also recognize that the larger the sample for which the light came up about 75% green, the more reasonable it is to believe that the light is not fair. For example, if 10,000 pushes resulted in 7500 green and 2500 red, students could be virtually certain that the dunking booth was not set as Mr. Rose had been told. If it came up 7 times green out of 10 pushes, they couldn't be as sure that the booth was not set up fairly.

Note: At this point, students have no precise way of judging these matters, but in fact, if the button works as Mr. Rose was told, and if it were pushed 20 times, there is only about a 2% chance of getting 15 or more dunks. The chance of 45 or more dunks in 60 tries is only about .00007, or 1 out of about 150,000, and even for as few as 100 pushes, the chance of getting dunked 75% or more of the time is less than 1 in a million.

How Different Is Really Different?

Intent

Students continue to explore the competing criteria of sample size and "degree of unusualness" in determining how much significance to attach to an "unusual" data set.

Mathematics

In this activity, students will find that sample size, numeric difference, and percentage difference are all important in evaluating data. They will also begin to consider the "expected" sample results that will be needed eventually for determining χ^2 and begin to develop a tabular format for recording those values.

Progression

Students work individually or in groups to compare data on three coins and try to decide which is most "suspicious." The subsequent discussion will help them organize their responses into charts comparing observed and expected numbers. Save the charts for future use.

Approximate Time

10 minutes for activity (at home or in class)

15 minutes for discussion

Classroom Organization

Individuals or groups, followed by whole-class discussion

Doing the Activity

Question 1a asks students to find the "expected" number of heads and tails if the coins were fair. As with the term **expected value** used in *The Game of Pig*, this terminology can be somewhat misleading. If a fair coin is flipped 100 times, we don't "expect" it to come up heads exactly 50 times. In fact, that will happen only about 8% of the time, although it is more likely than any other result. The meaning here, like the meaning of expected value, is the "long run" or "average" expectation. You may need to clarify this for students.

Discussing and Debriefing the Activity

Begin by having students report on Question 1. How did your expected numbers compare to the actual numbers for Alberto's coin? For Bernard's coin? For Cynthia's coin? They should give the expected numbers (the 50% values) and make some comparison of them with the actual results.

Suggest that it is easier to examine this information if it is well organized. How could you organize this information to make the comparison easier? The chart format described here will be very helpful in working on the more complex problems to come.

Set up three charts comparing the expected number of heads and tails if the coins were fair with the numbers actually observed. Leave room for recording chi-square values and related probabilities, to come later in the unit. Before the numbers are entered, the charts might look like this.

	Heads	Tails	Total
Alberto's coin	Expected: Observed:	Expected: Observed:	

	Heads	Tails	Total
Bernard's coin	Expected: Observed:	Expected: Observed:	

	Heads	Tails	Total
Cynthia's coin	Expected: Observed:	Expected: Observed:	

You might ask members of different groups to supply the various expected numbers. Be sure they can articulate that for a fair coin, the expected number is half the total number of flips. Then have students fill in the observed numbers. The completed charts should look something like this. Save these charts for later use.

	Heads	Tails	Total
Alberto's coin	Expected: 10 Observed: 14	Expected: 10 Observed: 6	20

	Heads	Tails	Total
Bernard's coin	Expected: 50 Observed: 55	Expected: 50 Observed: 45	100

	Heads	Tails	Total
Cynthia's coin	Expected: 500 Observed: 460	Expected: 500 Observed: 540	1000

For Question 2, take a straw vote and record next to each chart how many students voted for that coin as most suspicious. Then ask representatives of various groups to report on their reasoning. Why did you choose the coin you did? Students may not agree, and it is important to hear their explanations.

One perspective is to say, for example, that Alberto's coin has 4 heads more than expected while Bernard's coin has 5 more than expected. This might make Bernard's result seem "further off." On the other hand, Bernard's coin is 55% heads while Alberto's result is 70% heads, which makes Alberto's coin seem "further off." Similarly, the numeric difference between observed and expected values for Cynthia's coin is 40 flips, which is 4% of the total flips, while Bernard's coin is off by 5% (5 flips out of 100). Yet the much larger number of flips for Cynthia's coin makes that 4% "error" seem more significant than Bernard's 5%. (Statistically, it turns out that Cynthia's coin is the most suspicious and Bernard's coin is the least suspicious, but don't expect students to reach a consensus. What is important is students' reasoning about which coin is most suspicious.)

To reinforce the concept of **sampling fluctuation,** you might ask, Doesn't the data set show that *all* the coins are unfair? If necessary, point out that none of the coins came up 50% heads and 50% tails. Students should be able to articulate that this doesn't prove the coins are unfair and that a fair coin won't always produce perfectly balanced results. Look for an opportunity to use the term *sampling fluctuation* in this context.

Students should gradually come to recognize the need to look at two kinds of differences to measure how unusual a result is for a fair coin.

The *numeric* difference between expected and observed values

The *percentage* difference between expected and observed values

Key Questions

How did your expected numbers compare to the actual numbers for Alberto's coin? For Bernard's coin? For Cynthia's coin?

How could you organize this information to make the comparison easier?

Why did you choose the coin that you did?

Doesn't the data set show that *all* the coins are unfair?

Whose Is the Unfairest Die?

Intent

Students continue to explore the competing criteria of sample size and "degree of unusualness" in determining how much significance to attach to an unusual data set.

Mathematics

This activity is very similar to *How Different Is Really Different?* This time, though, the two amounts being counted—1s versus other rolls on a die—are not equally likely. Students evaluate hypotheses based on new data and intuitively evaluate which data set differs most from the expected values. They are also introduced to calculating expected results for a sample for which the probability is not 50%.

Progression

Students explore the activity individually and share results in a class discussion. As in *How Different Is Really Different?,* preserve the expected and observed results in chart form for later use.

Approximate Time

15 minutes for activity (at home or in class)

15 minutes for discussion

Classroom Organization

Individuals, followed by whole-class discussion

Doing the Activity

Be sure students recognize that the key probabilities involved in this situation are $\frac{1}{6}$ and $\frac{5}{6}$, not $\frac{1}{2}$ and $\frac{1}{2}$, bringing out that this is related to the fact that a die has six faces.

Discussing and Debriefing the Activity

For Question 1, have students report for the three dice, giving the expected numbers and comparing them with the observed numbers. In each case, the expected number of 1s is $\frac{1}{6}$ of the total number of rolls, and the expected number of other rolls is $\frac{5}{6}$ of the total. Be sure students recognize that these numbers are based on the assumption that the dice are fair.

As in *How Different Is Really Different?*, this information can be presented in charts, which might look like these. Again, emphasize that the expected numbers are based on the assumption that the dice are fair.

	1s	Other rolls	Total
Xavier's die	Expected: 5 Observed: 1	Expected: 25 Observed: 29	30

	1s	Other rolls	Total
Yarnelle's die	Expected: $16\frac{2}{3}$ Observed: 23	Expected: $83\frac{1}{3}$ Observed: 77	100

	1s	Other rolls	Total
Zeppa's die	Expected: $166\frac{2}{3}$ Observed: 178	Expected: $833\frac{1}{3}$ Observed: 822	1000

Save these charts for later use.

Have students give their responses to Question 2 about which die is the most suspicious as you tally the results. As with the coin problem, students may differ in their evaluation of the likelihood of unfairness of the dice. The important thing is for them to state their reasoning as clearly as possible.

Some students may be uncomfortable with an expected number that is not a whole number. If this issue comes up, ask, for example, What does it mean when you say the expected number for Yarnelle's die is $16\frac{2}{3}$? The best explanation probably relates to the familiar idea of "the average in the long run." While one will never get a fractional number in an individual experiment of rolling dice, the more one rolls a fair die, the closer the results will get to a theoretical average, which can be a fraction.

Key Question

What does it mean when you say the expected number for Yarnelle's die is $16\frac{2}{3}$?

Coin-Flip Graph

Intent

Students will identify the phenomenon of sampling fluctuation in a frequency bar graph of coin-flip results and determine that the distribution is approximately normal.

Mathematics

In this activity, students construct a frequency bar graph of coin-flip results. They observe that the "expected" result is not necessarily the result that occurs most often, reinforcing the concept of *sampling fluctuation*. The bar graph is used as the basis for discussing the idea of a distribution of coin-flip results, helping prepare students for development of the chi-square distribution later in the unit. Students also calculate percentages for various ranges of results, laying the foundation for understanding the chi-square probability table.

Progression

In preparation for this activity, students break into pairs, flip a coin 100 times, and record their results to form a class data set with at least 50 values. Each student records the class data. Students then work on the activity individually. The follow-up discussion brings out that sampling fluctuation should be expected and that, in fact, the "expected" results do not occur most of the time. In addition, similarities are noted between the coin-flip distribution and a normal distribution.

Approximate Time

15 minutes for class data gathering

25 minutes for activity (at home or in class)

15 minutes for discussion

Classroom Organization

Pairs, then individuals, followed by whole-class discussion

Materials

Coins (about 10 per pair of students)

Doing the Activity

In preparation for the activity, students need to accumulate data about the number of heads in a 100-flip experiment. It is suggested they do this in pairs. If this activity is assigned as homework, allow time beforehand for gathering the class data.

Each pair of students will need ten coins. They can flip the entire set at once and count the number of heads, repeating the process ten times for a total of 100 flips.

Have each pair do this 100-flip experiment several times, recording the number of heads for each 100-flip set, until the class has accumulated at least 50 sets of results altogether. You do not need to have every pair do the experiment the same number of times.

Have students record the class data, which they will use to make a frequency bar graph for the activity.

Record the results for your own reference as well. They will be used in this activity and again in the discussion of *Decisions with Deviation*. You may also want to refer to them when introducing *Random but Fair*.

Discussing and Debriefing the Activity

Have an individual or group present the frequency bar graph, in which the height of each bar shows how often a particular number of heads occurred. It might look something like this graph, which shows, for example, that the result of 48 heads (with 52 tails) occurred five times, out of fifty 100-flip sets.

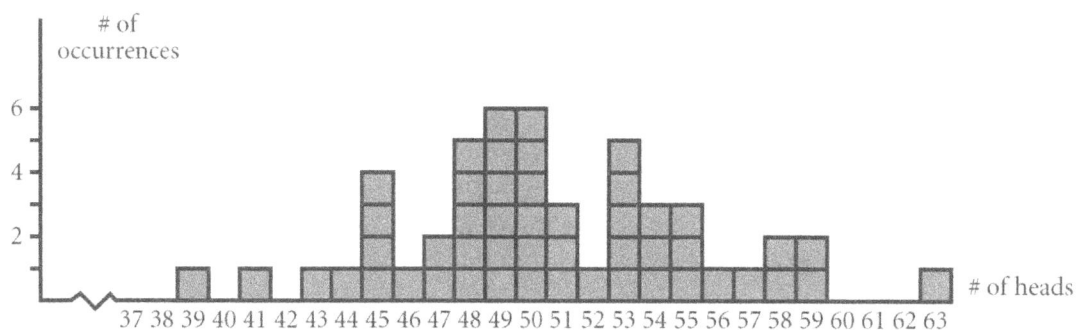

Ask students to comment on the results. They may be surprised at how widely spread the outcomes are. (The graph here shows the results of an actual calculator-based experiment. It is not especially unusual that a result as large as 63 heads occurred among the 50 experiments.)

Have other students give their answers to Questions 2 through 4.

Ask, **What do you think the mean of your data might be?** Students should say that it ought to be in the vicinity of 50, because about half the coin tosses should be heads. They should be able to confirm this visually by looking at the frequency bar graphs.

They should also realize that the mean of their experimental data will probably not be exactly 50. (They will use graphing calculators later to find the mean and the standard deviation.) Use the term **sampling fluctuation** in this context to explain why the experimental results usually don't match the probability of 50% heads, 50% tails.

Have students save these graphs (and the raw data) to look at again in the discussions of *Decisions with Deviation* and *Assigning Probabilities*.

Are Coin Flips Normally Distributed?

Ask students to imagine doing many, many sets of flips, with 100 flips in each set, and counting the number of heads in each set. Ask, Do you think a frequency bar graph of these data would be normally distributed? If needed, a general sketch of a bell-shaped curve will probably be a sufficient reminder for now about the **normal distribution.** It should be intuitively reasonable to students that at least "in the long run," the distribution might be close to normal.

In what ways is this distribution like the normal distribution? In particular, bring out that the bar graph of these probabilities has two important properties of the normal distribution.

It is symmetric. (For example, 58 heads is just as likely as 42 heads.)

It is largest at the center and gets smaller toward the ends. (For example, 50 heads is more likely than 51 heads, which is more likely than 52 heads, and so on. This may not be obvious, but it should make intuitive sense.)

Help students articulate ways in which the distribution is not normal. In what ways is this distribution different from the normal distribution? You might ask whether the result would ever have the smoothness of the normal curve. Students should recognize that the number of heads in a 100-flip experiment must be a whole number, so the distribution will always be a set of bars rather than the area under a smooth curve.

Tell students that the more coin flips per experiment, the closer the theoretical distribution will be to a normal distribution—close enough to make the theory of normal distributions useful.

Key Questions

What do you think the mean of your data might be?

In what ways is this distribution different from the normal distribution?

Do you think a frequency bar graph of these data would be normally distributed?

In what ways is this distribution like the normal distribution?

A Tool for Measuring Differences

Intent

These activities introduce the chi-square statistic as a tool for measuring the significance of data variations. Students learn how to compute this statistic and how to use it in deciding whether they should accept or reject a particular null hypothesis.

Mathematics

In *A Tool for Measuring Differences,* students review their work with **standard deviation** from the Year 1 unit *The Pit and the Pendulum* and apply the statistic to a variety of situations. Finding that standard deviation might not help with the situations from *Two Different Differences,* they look at numeric techniques for evaluating the "weirdness" of a data set. They learn a formula for measuring weirdness that combines percentage difference and numeric difference. They compute this **chi-square statistic,** or χ^2 statistic, for several situations, look for situations that would yield the same χ^2 statistic as a particular data set, and apply the χ^2 statistic to given situations. They also create a graph to show the distribution of the χ^2 statistic and look at how to assign probabilities based on the statistic.

Progression

These activities begin with a look back at standard deviation and how to use it. Students are then presented with two possible methods for measuring the "weirdness" of coin data and realize that any good measure must take sample size into account.

The formula for computing the chi-square statistic is introduced. As presented here, the formula applies to a situation in which data are compared with a theoretical probability. Students will learn how to adjust the formula for comparison of two populations later in the unit. In *How Does χ^2 Work?*, students find that the chi-square statistic combines aspects of numeric and percentage differences into a single formula. In *Measuring Weirdness with χ^2* and *χ^2 for Dice,* they apply this tool to data sets on coins and dice.

Different Flips begins a new phase of the work, in which students gather data about the chi-square statistic itself. In *Graphing the Difference,* they make a frequency bar graph of their data, and in *Assigning Probabilities* they convert those results into probabilities. They extend their data set on the chi-square statistic in *Random but Fair.* The goal of these activities is to offer insight into the probabilities associated with particular chi-square values. A crucial aspect of this work is recognizing that these probabilities are based on the **null hypothesis:** the chi-square statistic measures how unusual a given result would be if the null hypothesis were actually true.

Reference: Normal Distribution and Standard Deviation

Bacterial Culture

Decisions with Deviation

The Spoon or the Coin?

Measuring Weirdness

Drug Dragnet: Fair or Foul?

How Does χ^2 Work?

The Same χ^2

Measuring Weirdness with χ^2

χ^2 for Dice

Does Age Matter?

Different Flips

Graphing the Difference

Assigning Probabilities

Random but Fair

Reference: A χ^2 Probability Table

A Collection of Coins

POW 9: A Difference Investigation

Late in the Day

Reference: Normal Distribution and Standard Deviation

Intent

These reference pages summarize the basic facts about the **normal distribution** and **standard deviation.**

Mathematics

Reviewing previously learned attributes of standard deviation and the normal curve provides a good context for students' upcoming work with the **chi-square statistic.**

Progression

Open the discussion by talking about what students know about the normal distribution and then what they recall about how their work in the Year 1 unit *The Pit and the Pendulum* related to this topic. Then review the reference material with them.

Approximate Time

15 minutes

Classroom Organization

Whole-class discussion

Using the Reference Page

After using the coin-flip distribution from *Coin-Flip Graph* to briefly review the normal distribution, mention the Year 1 unit *The Pit and the Pendulum,* in which the normal distribution appeared for the first time in the IMP curriculum. Ask students how the normal distribution is related to their work from *The Pit and the Pendulum.* Here is a summary of the relevant activities.

Students measured the period of a "standard" pendulum and found that the measurement varied slightly as they repeated the experiment.

In compiling data from many repetitions of the experiment, students found the standard deviation of the data, which gave them a way of numerically quantifying the amount of expected "measurement variation."

Students did experiments on various "nonstandard" pendulums, in which they changed one of the basic parameters of the experiment. The parameters under consideration were the weight of the bob, the amplitude (angle) of the pendulum swing, and the length of the pendulum.

Based on the initial data on the standard pendulum, and especially on the standard deviation of that data set, students decided whether the changes they observed in the nonstandard pendulums were "real" changes or simply measurement variations.

Then ask students how they decided what mattered in those experiments. For the purposes of this discussion, bring out two key elements:

A review of the concept of measurement variation: There will be variation in the results of many kinds of experiments. You might draw an analogy to the variation students saw when they did the 100–coin-flip experiment many times.

A review of the *use* of that idea: When students changed the pendulum, the change in period was considered meaningful only if it was beyond the "typical" variation in their standard pendulum experiments.

Students should recall that their work with the normal distribution made it clear that some variation was normal and that a change should be considered significant only if it went beyond that typical variation.

If it hasn't yet come up, ask what tool students used for measuring this "typical" variation. This reference to standard deviation can serve as a lead-in to the next activity, *Bacterial Culture.* Tell students that they may want to refer to this reference material as they work on that activity.

Bacterial Culture

Intent

Students review **standard deviation** and **normal distribution** in a real-world context.

Mathematics

This activity focuses on how to use standard deviation to determine whether a difference is significant.

Progression

Students explore the activity in groups or individually and share findings in a class discussion.

Approximate Time

25 minutes

Classroom Organization

Groups or individuals, followed by whole-class discussion

Doing the Activity

Have students begin work on the activity.

Discussing and Debriefing the Activity

Let students from various groups give their responses to each question.

Questions 1 and 2 review fundamental facts about the normal curve and standard deviation. In the discussion of Question 1, point out, if presenters do not, that the change of concavity marks one standard deviation from the mean. For Question 2, the key idea is that approximately 95% of results are within two standard deviations of the mean.

For Question 3, the usual "there is no difference" language should be easy for students to apply to get the null hypothesis. But there may be several suggestions for the researchers' hypothesis. One reasonable idea is, "Infect-Away results in fewer bacteria than Bact-Out."

Make Question 4 the main focus of the discussion. Be sure students use standard deviation to explain their conclusions. Be looking for an explanation like, "The result is more than three standard deviations from the mean for Bact-Out. This result would be extremely unlikely if the two formulas were equally effective. So we should reject the null hypothesis and conclude that Infect-Away is probably more effective than Bact-Out."

If students don't refer to the null hypothesis in discussing this question, ask, How does the null hypothesis fit into the reasoning here? It's crucial that students recognize that they are finding the likelihood of getting a result like this *if the null hypothesis is true.*

Key Questions

What would you conclude from the experiment on Infect-Away?

How does the null hypothesis fit into the reasoning here?

Decisions with Deviation

Intent

This activity continues the review of standard deviation.

Mathematics

Students are asked to make some decisions using the only statistical tool they have, **standard deviation.**

Progression

Working individually or in groups, students analyze three situations using standard deviation. After discussing the solutions, the class compares the data from their previous coin-flip experiment to the normal distribution. Students observe that standard deviation is not easily applied to some of the other problems they have encountered in this unit. They will recognize a definite need for the **chi-square statistic** by the end of this activity.

Approximate Time

25 minutes for activity (at home or in class)

30 minutes for discussion

Classroom Organization

Individuals or groups, followed by whole-class discussion

Materials

Class coin-flip data from *Coin-Flip Graph*

Doing the Activity

Have students read over the three questions. Clarify the tasks if necessary.

Discussing and Debriefing the Activity

You might assign a couple of groups to present each of the questions.

For Question 1, students need to consider the fact that this is a "one-tail" problem. That is, they want to find the percentage of children at least one standard deviation above the mean. They will need to use the symmetry of the normal curve to answer this question. If needed, suggest they sketch the normal distribution for this set of heights, marking the mean and the first and second standard deviation points. Then they can find the percentages for various portions of the area.

A sketch of the distribution may also be useful for Question 2. Students should recognize that for a ball to "fail," it must be either more than two standard deviations below the mean or more than six standard deviations above the mean. Although the reference pages *Normal Distribution and Standard Deviation* do not provide information beyond two standard deviations, students will probably realize that hardly any balls will bounce more than 3 feet 2 inches, so they need only be concerned with those that bounce less than 2 feet 10 inches.

In addition to the numeric problem of finding the percentage of balls within a given range, there is the issue of whether the league requires *all* its balls to be within the given margin or is willing to allow an occasional ball to be outside the margin.

In Question 3, students start from a percentage and determine a range of values, rather than vice versa. If they sketch the distribution, you might ask, How much of the area is to the right of the mean? How much of the area is between the mean and the first standard deviation below the mean?

Back to the 100-Coin-Flip Data

Having reviewed the normal distribution and standard deviation over the course of the last couple of activities, the class can now examine how well the class coin-flip data from *Coin-Flip Graph* fit the normal distribution.

First have students find the mean and standard deviation for the class's 100–coin-flip data. Students can use their calculators to do the mechanical part of this, although you should review the mechanics of the computation. In particular, if you point out that the computation involves squaring the difference between the actual value and the mean, this may help make the formula for the chi-square statistic seem reasonable.

Ask students, Does the mean you get seem reasonable? They should have discussed previously that the theoretical mean is 50.

Students might also "eyeball" their frequency bar graphs to find out if the standard deviation seems reasonable. You might have them imagine a normal curve roughly fitted to their graphs, and review the idea that the concavity of the normal distribution changes at the places that are one standard deviation to the left and to the right of the mean.

The theoretical standard deviation is equal to 5, so the standard deviation for students' data should be close to that.

The class can now look more precisely at how well the data set fits the percentages from the normal distribution. (As pointed out earlier, the number of flips can only be a whole number, so this data set can't really be normally distributed.)

Have students find the percentage of their results that lie within one standard deviation of the mean and then compare this with the theoretical percentage (approximately 68%) for normal distributions. You might point out that they are using experimental values for both the mean and the standard deviation.

Then have students find the percentage of their results that lie within two standard deviations of the mean and again compare this with the theoretical percentage (approximately 95%) for normal distributions.

Students' percentages won't exactly match the theoretical percentages of the normal distribution, but they will probably be close enough to give the theory some credibility. You might ask, What would make our data set look more like a

normal distribution? Students will probably realize that they would need to increase the number of experiments.

A New Approach Needed

Tell students that although standard deviation might give fairly reasonable results for problems like the coin-flip experiment, it has two serious drawbacks for other kinds of problems.

The standard deviation might not be known. Even in the coin-flip problem, students got only an experimental estimate of the standard deviation, and one can't always do experiments like this to obtain data.

In some problems, one isn't measuring the variation in a single number (as students did with the number of heads). In particular, "To Market, to Market" (the soft drink problem from *Two Different Differences*) has a very different structure, and standard deviation doesn't apply there.

You may want to take a few minutes to review "To Market, to Market." The null hypothesis in that problem is, "Men and women are equally likely to prefer the new soft drink over the old one." But it doesn't say what percentage of either group will prefer the new soft drink. Students need some other way of measuring "how unusual" a given result is.

Key Questions

How much of the area is to the right of the mean? How much of the area is between the mean and the first standard deviation below the mean?

Does the mean you get seem reasonable?

What percentage of your results lie within one standard deviation of the mean? Within two standard deviations of the mean?

The Spoon or the Coin?

Intent

Students use experiments to estimate probabilities.

Mathematics

This activity differs from the coin-flip experiments in that students will probably have very little idea what to expect with spoons. As a result, they will have to consider how large an experiment they need to conduct to get reliable results. They will use their results to estimate the probability of a particular result.

Progression

Working individually, students conduct an experiment to estimate the probabilities of a spoon landing bowl up or bowl down when flipped. The main point that should come out of the follow-up discussion is that confidence in their results is directly related to the number of flips conducted.

Approximate Time

5 minutes for introduction

20 minutes for activity (at home or in class)

10 minutes for discussion

Classroom Organization

Individuals, followed by whole-class discussion

Materials

1 spoon for each student

Doing the Activity

You may want to introduce this activity and point out that spoons must be tossed into the air high enough to give them a chance to spin around.

Discussing and Debriefing the Activity

Begin the discussion by having two or three volunteers describe what they did and their results. Be sure they indicate how often they did the experiment and how often they obtained various results.

Some students may have flipped ten spoons at a time and looked at how often they got more than five or fewer than five landing bowl up. Others may have flipped a single spoon a large number of times and looked at the fraction that landed bowl up. Either approach is fine.

The essential issue is students' confidence in their conclusions. Although they do not yet have the mathematical tools to determine how many flips are needed for a given level of confidence, they should be able to articulate that the more flips one does, the more confident one can be about the results.

Measuring Weirdness

Intent

This activity will get students thinking about the specifics of how to measure "weirdness." In particular, it will help them to realize the limitations of two of the most natural approaches.

Mathematics

Students look at two techniques for evaluating the weirdness of data about coins and determine that neither numeric difference nor percentage difference is an adequate tool for this purpose.

Progression

Students work in groups to rank the three coins from *How Different Is Really Different?* for weirdness based on numeric difference and then on percentage difference. The follow-up discussion brings out that neither technique is adequate.

Approximate Time

40 minutes

Classroom Organization

Groups, followed by whole-class discussion

Doing the Activity

Have groups start right in on this activity, without any preliminary discussion.

Discussing and Debriefing the Activity

Let several students share a variety of examples to illustrate the difficulties with each method. Then, to illustrate the shortcomings of Alberto's method, you might compare these two cases:

Bernard's result of 55 heads and 45 tails

A result of 8 heads and 0 tails

Bernard's result has a numeric difference of 5 between the observed (55) and the expected (50) number of heads. The 8-heads/0-tails result has a numeric difference of only 4, so Alberto's method would rank that coin as less weird.

Ask, **Which of these two results *seems* weirder?** Point out that getting 8 heads in 8 flips seems weirder than 55 heads in 100 flips. In other words, a coin that yielded 8 heads out of 8 flips would seem less likely to be fair than a coin that yielded 55 heads out of 100 flips. Thus the coin that Alberto's method ranks as less weird seems intuitively weirder. In fact, this intuition is correct, because 55 heads in 100 flips will happen about 5% of the time, while 8 heads in 8 flips will happen only about 0.4% of the time. (Refer to "Or More?" for further discussion of this.)

To illustrate the shortcomings of Bernard's method, you might compare these two cases:

Alberto's result of 14 heads and 6 tails

A result of 130 heads and 70 tails

Alberto's result has a percentage difference of 20%, as 70% of his flips are heads, compared with 50% for a fair coin. The 130-heads/70-tails result has a percentage difference of only 15%, as 65% of the flips are heads, so Bernard's method would rank that coin as less weird. But the 130-heads/70-tails result will probably seem to be a weirder result to get from a fair coin than the 14-heads/6-tails result. Thus the coin that Bernard's method ranks as less weird actually seems weirder.

Although the activity asks students to focus on the shortcomings of each method, they should also realize that for a fixed sample size, either Alberto's or Bernard's method will tell which of two results is weirder. The problem is that neither method is adequate in comparing results for different sample sizes.

If students came up with suggestions for Question 3, let them share their ideas. Have the class discuss the validity of these ideas in terms of the kind of examples used for Questions 1 and 2.

Or More?

The preceding discussion is somewhat oversimplified. For example, in discussing how suspicious Bernard's coin is (55 heads and 45 tails), one should actually find the probability of getting 55 *or more* heads.

One way to explain this is to point out that if students are going to reject the null hypothesis in the case of Bernard's coin, they would have to reject it in any "weirder" cases as well. To illustrate with an extreme case, suppose one flipped a coin 1000 times and got exactly 500 heads. It turns out that the probability of obtaining this particular result from a fair coin is only about .025 (about 1 in 40), so this is a very unusual result. But it certainly wouldn't make one suspect the coin is unfair.

By comparison, a result of 4 heads out of 4 flips has a probability of about .06. Thus, it would be more likely for a fair coin to yield this result than the 500-heads/500-tails result. But one would probably be more suspicious of a coin that gave 4 heads out of 4 flips than of a coin that gave 500 heads out of 1000 flips.

This issue will be raised in *A χ^2 Probability Table* in the context of the frequency bar graphs of χ^2 statistics and will be explained there by analogy with students' work with standard deviation.

Key Questions

What's wrong with Alberto's method? What about Bernard's method?

Which of these two results seems weirder?

Drug Dragnet: Fair or Foul?

Intent

Students analyze a situation involving conditional probability.

Mathematics

The analysis used in the drug test problem involves the concept of *conditional probability.*

Progression

Students work individually to evaluate the confidence they can have that a positive drug test result is a true positive if the test is 98% accurate. The follow-up discussion formalizes the concept of conditional probability and brings out the social implications of a false positive result.

Approximate Time

5 minutes for introduction

20 minutes for activity (at home or in class)

15 minutes for discussion

Classroom Organization

Individuals, followed by whole-class discussion

Doing the Activity

Students may not know how to approach this problem. Point out the suggestion in Question 1 that they consider what might happen using a fairly large sample population.

Be sure students understand that *testing positive* means that the test indicates recent drug use. It will also help to introduce the terms *false positive* (a nonuser who tests positive), *true positive* (a user who tests positive), *false negative* (a user who tests negative), and *true negative* (a nonuser who tests negative).

Discussing and Debriefing the Activity

Try to get students to explain the problem using the terminology of *false positive, true positive,* and so on. Independent of the actual numbers, they are looking for the fraction $\dfrac{\text{number of true positives}}{\text{total number of positives}}$. They should realize that this fraction represents the probability that someone who tests positive is actually a drug user.

To better understand the situation, students can find the number of people in each of the four categories for a fairly large sample. For 10,000 people altogether, there would be 500 users and 9500 nonusers. Of the 500 users, 98% would test positive.

This gives 490 true positives. Of the 9500 nonusers, 2% would test positive. This gives 190 false positives. So, if someone tests positive, the probability that this person actually used drugs is $\frac{490}{680}$, or about 72%. This is considerably less than the term "98% reliability" might lead one to expect. Put another way, more than $\frac{1}{4}$, or about 28%, of people who test positive are actually not drug users.

Optional: An Area Approach

A diagram such as this may be helpful in understanding the various cases.

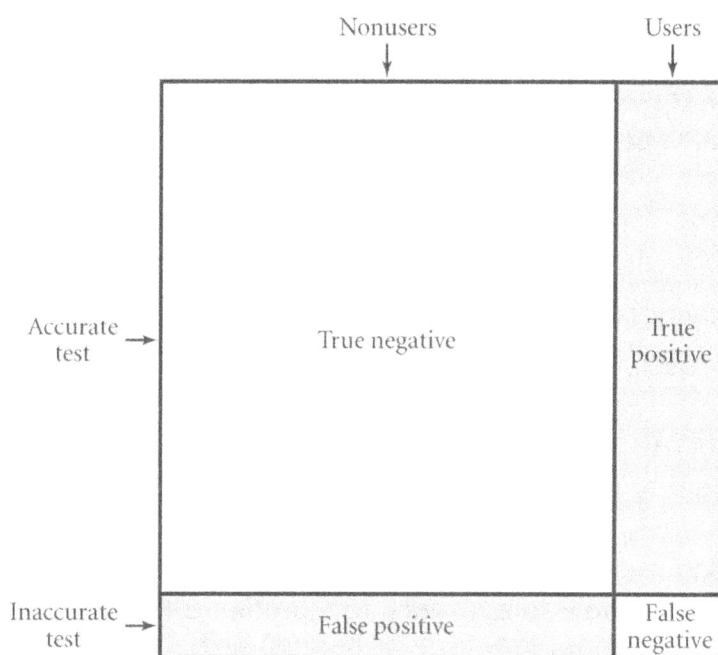

Although this diagram is not to scale, it does illustrate several things.

Nonusers heavily outnumber users. Nonusers actually represent 95% of the total.

For both nonusers and users, those testing "true" heavily outnumber those testing "false." Because the test is supposed to be 98% reliable, the "trues" are 98% of each group.

More than $\frac{1}{4}$ of those testing positive are actually nonusers.

As noted previously, Question 1 is concerned with the fraction $\frac{\text{true positives}}{\text{total positives}}$, which involves the two shaded regions of this diagram. Geometry can be used to show that the "true positive" area is about 2.5 times the "false positive" area, so the ratio of true positives to all positives is about 2.5 to 3.5, or $\frac{5}{7}$. The fraction $\frac{5}{7}$ is about 72%, which is the figure given previously for the percentage of people testing positive who are actually users.

Here is an outline of the area analysis.

The width of the "true positive" area is 2.5 times the height of the "false positive" area, because users are 5% of the total while false positives are 2% of nonusers.

The height of the "true positive" area and the width of the "false positive" area are both approximately equal to the side of the square representing the total area. One is 95% of that side; the other is 98%.

The area of the "true positive" section is thus about 2.5 times that of the "false positive" section.

Question 2: The Social Issues

Students may have a diversity of opinions about Question 2. Some may emphasize the public-safety importance of having certain workers be drug-free, while others may focus on the privacy issue.

Use the probabilities from Question 1 to bring out the likelihood of a "false positive" result, and make sure students realize how severely a false positive could affect someone. For example, such a finding might result in someone being unfairly fired from a job.

Students may suggest testing the positives a second time. If so, have groups calculate the probability that someone who tests positive a second time is indeed a true positive. (About 1% of those who test positive twice would actually be nonusers.)

Supplemental Activity

Smokers and Emphysema (extension) is a related activity also involving conditional probability.

How Does χ^2 Work?

Intent

Students compute the chi-square statistic for several situations.

Mathematics

In this activity, students learn to calculate the **chi-square statistic,** or χ^2 statistic. They will explore the statistic's sensitivity to sample size, numeric difference, and percentage difference and will find that the larger the χ^2 statistic, the more significant the deviation from the theoretical model. The development of the statistic is the key to Stage 6, hypothesis testing, in the stages of statistical investigation.

Progression

The teacher introduces the class to the χ^2 statistic, motivating their study of the formula by (1) comparing it with standard deviation and (2) presenting it as a blend of numeric difference and percentage difference. Students then work in groups to calculate χ^2 for a number of coin-flip results and observe how the statistic varies as the results get "weirder." The subsequent discussion includes a look at how the formula works.

Approximate Time

40 minutes

Classroom Organization

Teacher presentation, then groups, followed by whole-class discussion

Doing the Activity

Before introducing the χ^2 statistic, review the previous discussion of *Measuring Weirdness.* Ask students, What did you conclude about Alberto's and Bernard's methods for measuring weirdness? Review the fact that Alberto used numeric difference and Bernard used percentage difference, and summarize the discussion by noting that for different sample sizes, neither expression necessarily tells which of two results is weirder.

Tell students that there is an expression that works for different sample sizes. Statisticians have found that this expression can be used effectively to measure weirdness in many situations:

$$\frac{(\text{observed} - \text{expected})^2}{\text{expected}}$$

As this fraction expression is rather complex, it may help to provide some background. Ask students for general expressions for both numeric difference and percentage difference. They should recognize that numeric difference is

observed–expected

Getting a good expression for percentage difference is a bit harder. You might encourage students to proceed as follows.

Write observed percentage as

$$\frac{observed}{total}$$

Write expected percentage as

$$\frac{expected}{total}$$

The difference is

$$\frac{observed - expected}{total}$$

Thus the fraction

$$\frac{(observed - expected)^2}{expected}$$

incorporates both the numeric difference and the percentage difference, with *expected* replacing *total* in the denominator. Point out that for a fair coin flip, the denominator here, *expected,* will be exactly half the total number of flips. You might also point out that this last expression has the advantage of always being positive.

More precisely, the formula statisticians use is

$$\sum \frac{(observed - expected)^2}{expected}$$

That is, they compute the expression

$$\frac{(observed - expected)^2}{expected}$$

for each observed number and add the results.

For instance, in a fair-coin problem, there are two observed numbers: the number of heads and the number of tails. These two observed numbers are related, because their sum is the total number of flips, and the expected number for each is half the total number of flips.

Tell students that the number they get when they use this formula is called the **chi-square statistic.** (*Chi* is pronounced like "sky" without the "s.") Explain that chi is a letter in the Greek alphabet. A lowercase chi is written χ, and one often writes "χ^2 statistic" or just "χ^2" instead of "chi-square statistic."

As an example, calculate the χ^2 statistic for a sample of 50 coin flips of which 32 are heads. The expected/observed table would look like this.

	Heads	Tails	Total
50 coin flips	Expected: 25 Observed: 32	Expected: 25 Observed: 18	50

The χ^2 statistic is computed as

$$\chi^2 = \frac{(32-25)^2}{25} + \frac{(18-25)^2}{25} = \frac{7^2}{25} + \frac{(-7)^2}{25} = \frac{98}{25} = 3.92$$

The numeric value of 3.92 for χ^2 will have no intrinsic meaning for students at this point. It is similar to knowing that something is 3.92 standard deviations from the mean without knowing anything about the normal distribution.

In the activity, students will calculate the χ^2 statistic for coin-flip results that they make up, giving them some facility with the mechanics of the statistic. More importantly, they will use their results to examine whether this new statistic has the properties that were sought in *Measuring Weirdness* and that neither numeric difference nor percentage difference have.

If groups have trouble getting started on Question 1, have them simply choose a sample size, such as 50 or 100. If they need more encouragement, ask them to find the χ^2 statistic for the case in which the results match the expected numbers perfectly. For instance, have the group calculate the χ^2 statistic for a sample of 50 flips that yielded exactly 25 heads. The group should find that the χ^2 statistic in this situation is zero.

Then ask, What happens as the result gets further from "perfect"? They can examine a sequence of cases in which the number of heads for, suppose, 50 flips gradually increases from 25 to the maximum of 50.

Questions 2 and 3 are designed to help students realize that unlike Alberto's and Bernard's methods, the χ^2 statistic is sensitive to changes in sample size. They should conclude that the result that is intuitively weirder does indeed yield a larger χ^2 statistic.

Discussing and Debriefing the Activity

Have volunteers share their results for Question 1. Be sure someone calculates the χ^2 statistic for the situation in which the results fit the expected numbers perfectly so that everyone can note this special case. You might also include the case in which the results are as far off as possible. The class should notice that for a fixed sample size, as the results get weirder, the χ^2 statistic increases.

Have students present their χ^2 statistics and conclusions for Questions 2 and 3. Presumably they will agree that in each case, again, as the results get weirder, the χ^2 statistic increases. In Question 2, they should recognize the numeric difference of 10 as more significant for the smaller sample size. In Question 3, they should recognize a percentage difference of 25% as more significant for the larger sample size. (In Question 2a, the χ^2 statistics are 0.4 and 13.3, respectively; in Question 3a, the χ^2 statistics are 3.0 and 15.0, respectively.)

Explaining How the Formula Works

Ask students if they can apply basic knowledge of fractions to explain how the formula responds to the various changes by looking at three separate cases:

Fixed sample size

Fixed numeric difference

Fixed percentage difference

They may be able to give explanations for the various cases along these lines.

Fixed sample size: The greater the difference between observed and expected, the larger the χ^2 statistic. The numerator in each term gets larger while the denominator stays the same.

Fixed numeric difference between observed and expected: The smaller the sample size, the larger the χ^2 statistic. The numerator in each term stays the same while the denominator gets smaller.

Fixed percentage difference between observed and expected: The bigger the sample size, the larger the χ^2 statistic. Both the numerator and the denominator increase with sample size, but the numerator increases more rapidly.

For Question 4, let students speculate on the significance of a particular χ^2 statistic. At this point, they may say something along the lines of, "A χ^2 statistic of more than 10 would be pretty weird."

Key Questions

What did you conclude about Alberto's and Bernard's methods for measuring weirdness?

Why does the formula respond the way it does to these various changes?

What happens as the result gets further from "perfect"?

The Same χ^2

Intent

This activity gives students more experience with computing the χ^2 statistic.

Mathematics

Students' practice in calculating the χ^2 statistic reinforces the concepts that χ^2 increases with increasing weirdness and that the significance of a numeric difference varies with sample size. Students also begin to develop a sense of the significance of a particular χ^2 value.

Progression

Working on their own, students look for two situations that would yield the same χ^2 statistic as a given data set. The follow-up discussion explores the variation in percentage differences and numeric differences, for the same χ^2, as sample size increases. The role of the null hypothesis when using the χ^2 statistic is also clarified.

Approximate Time

30 minutes for activity (at home or in class)

10 minutes for discussion

Classroom Organization

Individuals, followed by whole-class discussion

Doing the Activity

This activity requires little or no introduction.

Discussing and Debriefing the Activity

Have volunteers share their ideas for Question 1. They will likely have predicted numbers less than 80. Their intuition should suggest that for this larger sample, it would take a smaller percentage difference from the expected to produce an equally weird result.

One way to explain this is to say that while a fair coin might occasionally give 40 heads in 50 flips, it would be unlikely to continue at that rate for another 50 flips. If the second set of 50 flips were more evenly split between heads and tails, the percentage of heads would decrease.

The χ^2 results should confirm this intuition. Students should have found that 71 heads out of 100 flips gives approximately the same χ^2 statistic as 40 heads out of 50 flips.

For Question 2, the case of 1000 flips, intuition should again tell students that a smaller percentage of heads would be needed to create an equally weird result. Their computations should result in roughly 567 heads out of 1000 flips.

Help students articulate what is going on here. As the sample size increases from 50 to 100 to 1000 in these three equally weird results, two things happen.

The percentage differences from the expected 50% heads decrease. The results go from 80% heads to 71% heads to 56.7% heads.

The numeric differences increase. The results go from "15 away from expected" to "21 away from expected" to "67 away from expected."

You might ask students, Do these trends make intuitive sense to you? Why or why not?

χ^2 *and the Null Hypothesis*

It may help to review the role of the null hypothesis when using the χ^2 statistic for problems of the sort being studied in this unit.

Try to get students to realize that they need to answer a question like this: How likely are you to get a result that far (or further) from the expected result if the null hypothesis is true? They need to know whether that difference could simply be due to sampling fluctuation.

By now students should sense that the χ^2 statistic will help them answer this question. It turns out that χ^2 gives an excellent approximation of this probability for the preceding situations, as well as for other types of situations.

Key Questions

Do these trends make intuitive sense to you? Why or why not?

What's the role of the null hypothesis in the use of the χ^2 statistic?

How likely are you to get a result that far (or further) from the expected result if the null hypothesis is true?

Supplemental Activity

Explaining χ^2 *Behavior* (extension) asks students to show how the formula for the χ^2 statistic reflects the effect of sample size.

Measuring Weirdness with χ^2

Intent

Students continue building their understanding of how the χ^2 statistic works, applying the statistic to situations previously considered only at an intuitive level.

Mathematics

Students validate χ^2 as a tool for measuring weirdness as they compare the χ^2 statistics for the previous coin-flip problems with their intuitive sense of the weirdness in each situation.

Progression

Working in groups, students reexamine the coin results from *Measuring Weirdness* and Roberto's brother's coin from *Two Different Differences* using the χ^2 statistic. After the follow-up discussion, save the results for later reference.

Approximate Time

40 minutes

Classroom Organization

Groups, followed by whole-class discussion

Materials

Charts from the discussion of *How Different Is Really Different?*

Doing the Activity

Remind students of the three coins they explored in *How Different Is Really Different?* and ask, Which of the three coins will have the highest χ^2 statistic? Record their predictions, which will likely match their ideas about which coin was the most suspicious, even if those suspicions were incorrect.

With these predictions recorded, have students work in groups on the activity, in which they will calculate the χ^2 statistic for each of those coins. You might ask groups who finish first to write up their arithmetic for others to review.

Discussing and Debriefing the Activity

Have presenters describe what they did to get the χ^2 statistics for the three coins. For convenience, here are the charts of the coin data.

	Heads	Tails	Total
Alberto's coin	Expected: 10 Observed: 14	Expected: 10 Observed: 6	20

	Heads	Tails	Total
Bernard's coin	Expected: 50 Observed: 55	Expected: 50 Observed: 45	100

	Heads	Tails	Total
Cynthia's coin	Expected: 500 Observed: 460	Expected: 500 Observed: 540	1000

Have students compare the χ^2 statistics with their earlier predictions. If the results don't match their predictions, it will be interesting to note whether the χ^2 numbers convince them to change their opinions about the relative suspiciousness of the coins.

Save the results for possible use in later discussions about the χ^2 statistic.

Why Are the Fractions the Same?

Students may notice that in each case, the two fractions are equal. For instance, with Bernard's coin, the fractions are

$$\frac{(55-50)^2}{50} \text{ and } \frac{(45-50)^2}{50}$$

Point this out if no student does, and ask, Why do the two fractions come out the same each time?

The numerators are the same because if "observed heads" is above the expected number, "observed tails" must be an equal amount below the expected number, and the squares of the differences are the same. This phenomenon will occur in every use of the χ^2 statistic that students encounter in this unit (unless they do supplemental activities involving more than one degree of freedom).

The denominators are the same only because the expected numbers for heads and tails are both half the total number of flips. If a problem involved, say, the number of 1s rolled on a die, the denominators would differ. For instance, refer to the discussion of χ^2 for Dice.

Question 3

Have a volunteer share how to calculate the χ^2 statistic for Roberto's brother's coin. The χ^2 statistic is approximately 21.3—very suspicious. Students should sense by now that a χ^2 statistic this large is very unusual for a fair coin, though they don't yet know how to attach a probability to this number.

Which Method Is Best?

Ask the class, *What methods have you examined for measuring weirdness? Which of these methods seems best to you?*

Review the two methods proposed in Measuring Weirdness for measuring the weirdness of a coin-flip result: Alberto's method, using numeric difference, and Bernard's method, using percentage difference. If Measuring Weirdness with χ^2 accomplished its purpose, students will realize that only the χ^2 statistic is sensitive to sample size, which makes it the best of the three methods.

Tell students that because of its adaptability to different sample sizes, the χ^2 statistic is generally recognized as an excellent tool for situations like those in Two Different Differences.

Key Questions

Which of the three coins will have the highest χ^2 statistic?

Why do the two fractions come out the same each time?

What methods have you examined for measuring weirdness?

Which of these methods seems best to you?

χ^2 for Dice

Intent

Students apply the χ^2 statistic in a context for which the outcomes are not equally likely.

Mathematics

Students return to the dice data from *Whose Is the Unfairest Die?* to determine how well the χ^2 statistic supports their intuition about the weirdness of those results. In the process, they practice finding the χ^2 statistic, which will help them do the calculations reasonably smoothly. They strengthen their sense that a larger χ^2 statistic indicates a more suspicious result. In addition, the work emphasizes that the χ^2 statistic is useful for situations other than those in which the outcomes are equally likely.

Progression

Students work individually to calculate the χ^2 statistic for each of the three dice from *Whose Is the Unfairest Die?* and compare the results to their intuition about which set of results is the weirdest. The follow-up discussion solidifies the idea that the larger the χ^2 statistic, the weirder the result.

Approximate Time

20 minutes for activity (at home or in class)

10 minutes for discussion

Classroom Organization

Individuals, followed by whole-class discussion

Materials

Charts of dice results from *Whose Is the Unfairest Die?*

Doing the Activity

You may want to remind students that the expected number of 1s is not 50%.

Discussing and Debriefing the Activity

Have students review the charts from *Whose Is the Unfairest Die?*, and return specifically to what they thought then about which die is most suspicious. For convenience, here are the charts of the data.

	1s	Other rolls	Total
Xavier's die	Expected: 5 Observed: 1	Expected: 25 Observed: 29	30

	1s	Other rolls	Total
Yarnelle's die	Expected: $16\frac{2}{3}$ Observed: 23	Expected: $83\frac{1}{3}$ Observed: 77	100

	1s	Other rolls	Total
Zeppa's die	Expected: $166\frac{2}{3}$ Observed: 178	Expected: $833\frac{1}{3}$ Observed: 822	1000

Have students report the χ^2 statistic for each of the three dice. Save these statistics for future reference. For Xavier's die, the χ^2 statistic is found from the expression $\dfrac{(1-5)^2}{5} + \dfrac{(29-25)^2}{25}$, which equals 3.84. For Yarnelle's die, the χ^2 statistic is 2.89, and for Zeppa's die, the χ^2 statistic is 0.92 (given to the nearest hundredth).

Students should recognize once again that the larger the χ^2 statistic, the less likely the situation fits the null hypothesis. Thus Xavier's and Yarnelle's dice are about equally suspicious, and Zeppa's die is least suspicious.

Interpreting χ^2

By now students should be developing a clear intuitive sense of what it means to get a large χ^2 statistic, but they may still have trouble putting it into words.

Offer practice with this idea by asking, What is the significance of a big χ^2 statistic? What does this have to do with the null hypothesis? Help students articulate that it means the result they obtained would have been unlikely if the null hypothesis had been true. In other words, the bigger the χ^2 statistic, the more justified one is in rejecting the null hypothesis.

What does this mean in terms of the coin and die problems? In these cases, the size of the χ^2 statistic measures how weird it would be for a fair coin or die to give a result that weird (or weirder).

How is the χ^2 statistic similar to standard deviation? Students should recognize that both statistics can be used to measure how far a given result is from the mean. For example, saying that Yarnelle's die results give a χ^2 statistic of 2.89 is similar to saying that her results are a particluar number of standard deviations from the mean for 100 rolls.

Tell students they will soon begin to figure out how big is "big"—how big the χ^2 statistic should be before they can safely rule out the idea that the discrepancy is just a sampling fluctuation and reject the null hypothesis.

Key Questions

What is the significance of a big χ^2 statistic? What does this have to do with the null hypothesis?

What does this mean in terms of the coin and die problems?

How is the χ^2 statistic similar to standard deviation?

Does Age Matter?

Intent

Students continue to develop an understanding of how the χ^2 statistic works as they apply it to a real-world situation.

Mathematics

This activity puts the issue of the significance of a deviation from expectation into a decision-making context. Students begin to notice more of the stages of statistical investigation working together as they form a hypothesis and a null hypothesis and evaluate data from an experiment using the χ^2 statistic. They also realize that while χ^2 is a useful decision-making tool, it does not remove all uncertainty.

Progression

Because of the questions that are likely to arise, this activity is best suited for classwork rather than homework. Given some data on two populations, students work in groups or on their own to write hypotheses, calculate the χ^2 statistic, and make a decision. When they reach Question 3 (calculating χ^2), it will likely be necessary for the teacher to clarify that the situation involves a theoretical model, even though it compares two populations. Comparison with coin-flip data yielding the same χ^2 statistic may help students evaluate their results.

Approximate Time

40 minutes

Classroom Organization

Individuals or groups, followed by whole-class discussion

Doing the Activity

Read through the activity as a class, and then have students, on their own or in groups, get to work.

Discussing and Debriefing the Activity

You will probably need to bring students together after they have worked through Questions 1 and 2, as they will have trouble with Question 3 unless they have phrased the problem properly. The following discussion of Questions 1 and 2 will help students establish the proper framework for setting up Question 3.

Begin the discussion with a clarification of the null hypothesis and Clementina's hypothesis. If students have difficulty with the null hypothesis, ask, What might you think about the tipping habits of adults and high school students if you had no evidence to go on? You might remind them that a null hypothesis often begins, "There is no difference . . ." Students should reply that the null hypothesis should be something like, "There is no difference between high school students and adults with respect to tipping."

Students might say that even without data, they would expect the two groups to tip differently. Point out that they are, in fact, using informal evidence in making that judgment—namely, evidence based on their personal experience with the two groups, though not necessarily data about tipping in particular.

You may want to use this occasion to review the idea of data snooping as a method of coming up with hypotheses. You might also point out that our everyday experiences often provide "data" that we can use in formulating hypotheses, without formal "snooping."

Ask students to state more precisely what they mean by "with respect to tipping." Help them to realize that they are focusing on the percentage of people within each group who are good tippers (whatever that means).

According to the activity, Clementina doesn't think the null hypothesis is true. Her hypothesis could be stated as, "A smaller percentage of high school students than adults are good tippers."

Comparing Populations or Comparison with a Model?

As just described, this problem is similar to the soft drink problem "To Market, to Market" from *Two Different Differences.* The situation here is somewhat different, though, because Clementina has a large amount of data (from her mother) indicating that 70% of adults are good tippers. Therefore, this problem is really more like Roberto's coin-flip problem in *A Suspicious Coin.* In both cases, an unknown (the brother's coin or the high school tippers) is being compared with a known (the fair coin or the adult tippers).

How could you restate the null hypothesis using this perspective? Students might come up with something like, "70% of high school students are good tippers." If they try to use "there is no difference" language, they might state the null hypothesis as something like, "There is no difference, with respect to tipping, between the population of high school students and a population with 70% good tippers."

How could you restate Clementina's hypothesis using this perspective? It might be stated as, "Fewer than 70% of high school students are good tippers."

Computing the χ^2 Statistic

Have students present the calculation of the χ^2 statistic for Clementina, which should be straightforward once the null hypothesis is clarified. If they make a chart as they have done before, it will look like this.

	Good Tips	Poor Tips	Total
Clementina's customers	Expected: 36.4 Observed: 30	Expected: 15.6 Observed: 22	52

Students should be able to compute the χ^2 statistic as approximately 3.75. The figure 36.4 is 70% of the total of 52 customers; 15.6 represents the remaining 30% of the customers.

What is this χ^2 statistic measuring in terms of the null hypothesis? The null hypothesis is that 70% of high school students are good tippers. Therefore, the χ^2 statistic of 3.75 roughly measures how unusual it would be to get a sample of 30 good tips and 22 bad tips if the "overall population" (tips from all high school students) consisted of 70% good tips. More precisely, χ^2 deals with how unusual it would be to get a result *at least this far* from the expected value.

To give students a more intuitive frame of reference for evaluating this result, have them figure out the answer to this question based on a null hypothesis of a fair coin: **How many heads out of 100 coin flips would also give a χ^2 statistic of approximately 3.75?** They should find that the closest whole-number result is 60 heads (or 40 heads). Clementina getting 30 good tips out of a sample of 52, taken from a population that gives 70% good tips, is about as unusual as getting 60 heads when flipping a fair coin 100 times.

What Should Clementina Do?

Focus the end of the discussion on the practical implications of the data. Based on their work, students will probably think there is a good chance that Clementina will earn more serving adult customers but that it isn't a sure thing.

Some students will likely say that job satisfaction is the most important factor, regardless of the monetary compensation, raising the point Clementina has to weigh the chance to earn more against the possibility that the job might not be as fun.

Key Questions

What might you think about the tipping habits of adults and high school students if you had no evidence to go on?

How could you restate the null hypothesis using this perspective?

How could you restate Clementina's hypothesis using this perspective?

What is this χ^2 statistic measuring in terms of the null hypothesis?

How many heads out of 100 coin flips would also give a χ^2 statistic of approximately 3.75?

Different Flips

Intent
Students gather data about the χ^2 statistic based on coin flips.

Mathematics
This activity gives students more practice calculating the χ^2 statistic. The pooled class data will provide insight into the probability of obtaining a χ^2 statistic that is large or larger than the calculated one when the null hypothesis is known to be true (in this case, when the coin is fair).

Progression
The activity is introduced with a discussion of the need for being able to associate probabilities with χ^2 statistics and of a plan for developing such an association. On their own, students perform three coin-toss experiments, each with a different number of tosses, and calculate the χ^2 statistic for each case; the whole class then pools their data. Students will make a frequency bar graph of the pooled data in *Graphing the Difference* and associate probabilities with varying χ^2 statistics in *Assigning Probabilities.*

Approximate Time
10 minutes for introduction

25 minutes for activity (at home or in class)

10 minutes for discussion

Classroom Organization
Individuals, then groups, followed by whole-class discussion

Materials
Coins

Doing the Activity
Students have observed that the larger the χ^2 statistic, the more justified one is in rejecting the null hypothesis. Point out that they now need a method for judging how big a "big" χ^2 statistic is. They need to know how unusual it is to get a particular χ^2 statistic when the null hypothesis is true in order to decide whether a particular χ^2 result justifies rejecting the null hypothesis.

Ask students to conduct their coin-flip experiments with care so that the class will have reliable data with which to assess χ^2 statistics.

Discussing and Debriefing the Activity

Have students work in groups to combine their data, perhaps putting results from 0 to 0.5 in one grouping, from 0.5 to 1.0 in the next grouping, and so on. Fifteen groupings (up to $\chi^2 = 7.5$) will probably cover all the responses.

Groups can then report their totals to the class. Have group representatives mark their totals on a transparency, or have groups report to you while you record the data.

You might have students double-check any χ^2 statistics greater than 5 to be sure they weren't the result of arithmetic error. In a class with 30 students each calculating three χ^2 statistics, you can expect only two or three such large values.

Graphing the Difference

Intent

Students will use the real data they collected to find various chi-squares and note *how* often they occur. The probabilities will give meaning to the chi-square likeliness so students can make sense of and use the chi-square probability table.

Mathematics

Students will group the data from their coin-flip experiments to construct frequency bar graphs, which they can then use to associate probabilities with χ^2 statistics.

Progression

Students work on their own to make frequency bar graphs of the pooled class data from *Different Flips* and answer some simple questions using their graphs. The class discussion will explore the fact that the χ^2 statistic is not normally distributed and review the use of a frequency bar graph to find probabilities. The frequency bar graphs will be used in the activity *Assigning Probabilities.*

Approximate Time

30 minutes

Classroom Organization

Individuals, followed by whole-class discussion

Materials

Class data from *Different Flips*

Doing the Activity

Before students begin work, you may want to mention that they will need the frequency bar graphs they will be creating in the activity *Assigning Probabilities.*

Discussing and Debriefing the Activity

Begin the discussion by asking the class, Do you think χ^2 statistics are normally distributed? Then ask what it is that makes students think the distribution isn't normal. They should point out, among other things, that the curve is not symmetrical.

The answers to Questions 2 to 4 will depend on the particular results that students got in *Different Flips,* but the general theory, as expressed in *A χ^2 Probability Table,* suggests these results.

Question 2: The percentage should be about 32%.

Question 3: The probability should be about .08.

Question 4: The probability should be about .95.

Key Question

Do you think χ^2 statistics are normally distributed?

Assigning Probabilities

Intent

Students explore how to assign probabilities based on the χ^2 statistic.

Mathematics

Students convert a graph of χ^2 statistics from the theoretical-model case into a table of probabilities.

Progression

Working individually, students use the frequency bar graph from *Graphing the Difference* and the χ^2 statistic for Bernard's coin from *Measuring Weirdness with* χ^2 to estimate the probability of getting a χ^2 statistic of 1, assuming the null hypothesis is true. They will extend their data set on the chi-square statistic in *Random but Fair.*

Approximate Time

30 minutes

Classroom Organization

Individuals, followed by whole-class discussion

Materials

Frequency bar graphs created in *Graphing the Difference* (for the activity)

Frequency bar graphs created in *Coin-Flip Graph* (for the discussion)

Doing the Activity

This activity requires little or no introduction.

Discussing and Debriefing the Activity

Ask someone to explain how he or she estimated the probability described in Question 1c. This explanation should be similar to students' reasoning on Questions 2 to 4 of *Graphing the Difference.* The answer will depend on the data students gathered. Bernard's coin has a χ^2 statistic of 1.0, and the theoretical probability of getting a χ^2 statistic that large or larger is about .32.

A comparison of the result for Bernard's coin with students' data from *Coin-Flip Graph* may give the χ^2 statistic some credibility. Review the fact that the χ^2 statistic for Bernard's coin is 1.0, and ask, What coin-flip results would give χ^2 statistics greater than or equal to 1.0? Students should realize that any result that falls "as far or further" from the mean as Bernard's coin will do so. So they should count cases in which there are 55 or more heads and cases in which there are 45 or fewer heads. Their findings should be roughly the same as the proportion of χ^2 statistics in their frequency bar graphs that are 1.0 or higher.

Key Question

What coin-flip results would give χ^2 statistics greater than or equal to 1.0?

Random but Fair

Intent

Students gather data about the χ^2 statistic based on digits randomly generated to simulate a statement that 70% of adults are good tippers.

Mathematics

Students will convert a graph of χ^2 statistics from the theoretical-model case to a table of probabilities and note why the χ^2 probability table gives probabilities for getting a given χ^2 statistic or larger.

Progression

Working in pairs, students conduct a simulation using a random number generator and find the χ^2 statistic for each experiment. The class makes a frequency bar graph of the results and notices that it generally resembles the graph from *Graphing the Difference.* The ensuing discussion will help students make the transition to a "smooth" graph of the theoretical χ^2 distribution.

Approximate Time

30 minutes

Classroom Organization

Pairs, followed by whole-class collation of data, graph construction, and discussion

Materials

Chart graph paper

χ^2 *Distribution Graph* blackline master (transparency)

Doing the Activity

Have students read the introduction, and then review how the experiment will simulate Clementina's null hypothesis. You may need to review both the concept of a random number generator and the mechanics of generating random numbers with the calculator.

It may be helpful to do one experiment as a class, generating the 40 values as described in Question 1 and computing the χ^2 statistic. As with *Different Flips,* the reason to vary the number of trials is to avoid getting the same χ^2 statistic over and over.

While students are generating results and statistics, you can set up a frequency bar graph with a scale identical to that used in *Graphing the Difference.* As students generate each set of random numbers and compute the χ^2 statistics, they can build the class frequency bar graph, which they will compare to the graph made in the previous activity.

Discussing and Debriefing the Activity

Once the graph is complete, ask the class to compare it to the graph from *Graphing the Difference.* The graphs should have similar shapes. Tell students that if they had calculated enough χ^2 statistics in both cases, with varying sample sizes, the two graphs would be nearly identical.

Also explain that although the graphs for the two situations are not identical, there is a "standard" graph, called the χ^2 *distribution curve,* that can be used to approximate them both. This standard graph can also be used to approximate similar graphs for all the other situations to which students will be applying the χ^2 method.

From Experimental Data to Theory

Ask students to imagine doing many experiments, using many different values for the number of flips or the number of random numbers generated, and then "smoothing out" the resulting frequency bar graph into a curve. The graph would look something like the one shown here. A larger version of this graph is given on the blackline master χ^2 *Distribution Curve.*

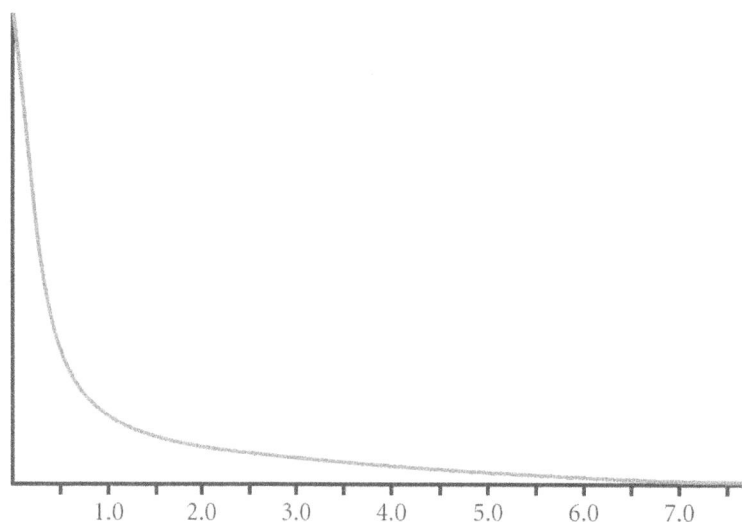

Note: If you choose a particular sample size and make a frequency bar graph of the outcomes, you will notice a "bump" near the beginning of the graph; the graph will seem to start by going up before going down as the χ^2 values increase. For example, if you count the number of heads in 40 coin flips, you are more likely to be off by 1 from the expected number of heads than to get 20 exactly, because both 19 heads and 21 heads are off by 1, which makes a χ^2 statistic of 0.1 more likely than a χ^2 statistic of 0. But when you consider all possible sample sizes and treat the χ^2 statistic as a continuous phenomenon, this "bump" disappears.

Have students articulate what this graph represents; suggest an analogy with the normal curve if needed. **What does this curve represent in terms of area and probability?** Students should be able to explain that the area to the right of any vertical line represents the probability of getting a χ^2 statistic that large or larger.

Use an example to illustrate this idea. For instance, students should realize that about one-third of the area under the curve is to the right of $\chi^2 = 1.0$ and that this area represents the probability that a fair coin would give a χ^2 statistic of 1.0 or more. This is consistent with their data that about one-third of their 100–coin-flip experiments gave results that would yield χ^2 statistics of 1.0 or more—namely, results of either 55 or more heads or 45 or fewer heads.

It may help students to think of the area under the graph as a "rug" with two colors separated by the χ^2 statistic they are assessing. For instance, in the coin-flip example just discussed, the area where χ^2 is more than 1.0 would be one color and the area where χ^2 is less than 1.0 would be another color.

Key Questions

How does the activity simulate Clementina's null hypothesis?

What does this curve represent in terms of area and probability?

Reference: A χ^2 Probability Table

Intent

Students examine a table showing the probability of getting a χ^2 statistic of a given value or higher.

Mathematics

The χ^2 probability table gives the likelihood of getting a χ^2 statistic of a given value or larger when the null hypothesis is true. These probabilities grow more accurate as the size of the experiment increases, and they provide reasonable approximations if all the expected numbers are at least 5.

Progression

In a class discussion, students examine a χ^2 probability table and discuss its use.

Approximate Time

10 minutes

Classroom Organization

Whole class

Materials

χ^2 *Distribution Graph* blackline master (transparency)

Using the Reference Page

Display the χ^2 distribution curve, and tell students that it's convenient to have detailed information about the probabilities from the curve in table form so they don't have to estimate areas every time they come across a χ^2 statistic. Have them turn to the reference pages in their books. Explain that the probabilities in this "official" table are only approximations of the probabilities in any given situation. The larger the size of a particular experiment, the better this approximation will be.

Explain that to ensure this approximation is "reasonably good," they should observe this guideline.

The χ^2 probability table should be used only if all the expected numbers are at least 5.

Point out that the χ^2 probability table refers to the probability of getting a χ^2 statistic "that large or larger," and ask, Why does the table refer to the probability of getting a χ^2 that large or larger and not a χ^2 of a particular size? The crucial observation is that if a given result would lead one to reject the null hypothesis, a "weirder" result would also lead one to do so. Thus the class of "rejecters" consists of all results with a χ^2 statistic of a given value or higher.

You can build on students' previous experience with standard deviation to clarify this explanation. In discussing standard deviation, students did not look at the probability of getting a result that was exactly one standard deviation from the mean. Instead they used the fact that about 68% of results were *within* one standard deviation of the mean. Similarly, they learned that about 2.5% of all results were *more than* two standard deviations above the mean. By analogy, it makes sense to talk about the probability of getting a χ^2 statistic that is *less than* or *more than* a given value, rather than the probability of getting a specific value.

Key Question

Why does the table refer to the probability of getting a χ^2 that large or larger and not a χ^2 of a particular size?

Supplemental Activity

Completing the Table (extension) explores the idea of linear interpolation in the context of the χ^2 probability table.

A Collection of Coins

Intent

This is another opportunity for students to observe how the χ^2 statistic responds to changes in experimental results.

Mathematics

Students will further study the relationship between sample size and the corresponding chi-square statistics.

Progression

Working individually, students practice using the χ^2 probability table and then make up samples with larger or smaller χ^2 statistics, but with the same numeric difference or number of trials. The subsequent discussion brings out the fact that the χ^2 statistic can give only probabilistic information, not an absolute answer about whether a particular null hypothesis is true.

Approximate Time

30 minutes for activity (at home or in class)

10 minutes for discussion

Classroom Organization

Individuals, followed by whole-class discussion

Materials

χ^2 *Distribution Graph* blackline master (transparency)

Doing the Activity

This activity requires little or no introduction.

Discussing and Debriefing the Activity

You might have students compare answers in their groups and then ask for volunteers to respond to each question.

You may get various responses and explanations for Question 1. One good explanation is that Al realized that Betty had flipped so many times that it was extremely unlikely her results were due to a sampling quirk. For instance, suppose Betty had flipped her coin 200 times and gotten only 80 heads. The probability of being this far from 50% is less than .005 ($\chi^2 = 8$). If Al had flipped his own coin only 4 times and gotten 1 head, he could easily ignore his 25% result.

You can use this problem as another opportunity to talk about sampling fluctuation by asking, **Isn't Al's coin also unfair, because he didn't get 50% heads?** By now students should be able to state clearly that one doesn't expect random events like coin flips to follow the probabilities perfectly but that they will approach the probabilities "in the long run."

As students give the χ^2 statistics for Question 2, ask them to refer to the graph of the χ^2 distribution and show where the results lie. **Where does each of these results fit in the χ^2 distribution?** This should bring out that none of these results is particularly unusual for a fair coin. The probabilities are approximately these:

Question 2a: .20 (χ^2 = 1.67)

Question 2b: .35 (χ^2 = 0.87)

Question 2c: .51 (χ^2 = 0.45)

Students may want to know how to get the probability for a χ^2 statistic that is not on the table. For instance, in Question 2a, the χ^2 statistic of 1.67 is between 1.6 and 1.8. At a minimum, students should recognize that the associated probability is between .2059 and .1797 and is closer to .2059 (because 1.67 is closer to 1.6 than to 1.8). For the purposes of this unit, it's fine if they simply estimate based on that reasoning. If they are interested in more details, you can refer them to the supplemental activity *Completing the Table,* which presents interpolation using proportional reasoning.

Questions 3 and 4 review the issues students examined in *How Does χ^2 Work?* For Question 3, students should notice that if they increase the sample size while keeping the same numeric difference, they will get a smaller χ^2 statistic. For example, a result of 55 heads and 65 tails will give a smaller χ^2 statistic than a result of 25 heads and 35 tails. In Question 4, to get a larger χ^2 statistic from the same sample size, one needs to make the results further from the expected results. For example, a result of 900 heads and 1120 tails will give a larger χ^2 statistic than a result of 995 heads and 1025 tails.

Will χ^2 Resolve the Questions?

It's important for students to recognize that the χ^2 statistic can give only probabilistic information rather than an absolute answer about whether a particular null hypothesis is true.

You can raise this issue by asking, **Now can you tell if Cynthia's coin is fair?** You might pose similar questions about other situations students have looked at. Bring out that χ^2 statistics do not give guarantees. Cynthia's coin (in the activity *How Different Is Really Different?*) and Xavier's die (in the activity *Whose Is the Unfairest Die?*) are the most suspicious, because their results would be the least likely for a fair coin or die, but one can't be certain based on the χ^2 statistic.

Students may be disappointed that the answers are not more definitive. Help them to understand that there is no way to know absolutely whether a coin or die is fair simply by doing a set of flips or rolls. They can only speak of probabilities—that is, they can only determine how likely they are to obtain a given result if a coin or die were fair. You might connect this uncertainty with the "confidence scales" students used in their work on *Two Different Differences.* There is no "magic" χ^2 statistic beyond which one should always reject the null hypothesis, though students may ultimately choose a particular cutoff point for their decisions.

Key Questions

Isn't Al's coin also unfair, because he didn't get 50% heads?

Where does each of these results fit in the χ^2 distribution?

Now can you tell if Cynthia's coin is fair?

POW 9: A Difference Investigation

Intent

This POW, the culmination of the unit, will give students an opportunity to synthesize many of the concepts and skills they are learning.

Mathematics

Students will conduct an investigation that applies all of the stages of statistical investigation studied in this unit, including application of the chi-square statistic.

Progression

The work for this POW is divided into seven phases that will need to be spread out over eight or nine days. Don't let this POW run over into the next unit, as it serves as a wonderful review for the unit assessments.

Phase 1: Introduce the POW and ask students to think about selecting a partner. Establish completion dates for the milestones.

Phase 2: Have students hand in their partner selections.

Phase 3: Spend some time discussing what makes an appropriate hypothesis and null hypothesis. This will build on students' further experiences using the χ^2 statistic. This is best scheduled immediately after students have completed *Big and Strong*.

Phase 4: Collect from each team a statement of their hypothesis and null hypothesis, as well as a plan for collecting data. This is best scheduled for the day after Phase 3 to allow students time to consider their hypotheses.

Phase 5: Return statements to the teams, with any necessary feedback on their hypotheses and plans, on the day after Phase 4 so teams have time to make any necessary changes. Have teams start collecting data.

Phase 6: Have teams work in class on completing their reports, particularly incorporating the use of the χ^2 statistic. Schedule this for at least two days after Phase 5, if possible, to allow sufficient time for collecting data.

Phase 7: Collect reports and begin presentations. This may be scheduled for the day after the in-class work of Phase 6.

Approximate Time

2–to 4 hours for activity, in addition to the following time in class:

20 minutes for Phase 1

5 minutes for Phase 2

10 minutes for Phase 3

45 minutes for Phase 6

85 minutes for Phase 7

Classroom Organization
Pairs and whole class for discussion and presentations

Materials
Schedule for POW 9: A Difference Investigation blackline master (optional; 1 per student and 1 transparency)

Doing the Activity
For the presentations in Phase 7, you may want to invite special guests, such as administrators, other teachers, parents, and students in IMP Years 3 and 4 and interested students in other classes. If so, it's probably wise to begin planning invitations soon.

Phase 1: Introduce the POW

Read the POW as a class and then review the phases described previously in "Progression." You may want to provide each student with a copy of *Schedule for POW 9: A Difference Investigation* and use a transparency of the master to fill in the timetable together, following the guidelines for establishing those dates under "Progression." Let students know about time they will have in class to work on the POW.

Acknowledge that students do not yet have the skills they need to analyze the data they will collect, but assure them that the activities between now and then will prepare them to do so.

Tell students they need to choose their partners by the date you have set and to begin thinking about a topic for their hypotheses. You may want to set limits on what you consider appropriate subject matter for these investigations.

Warn students that they may need another day or two to understand exactly how to formulate their hypotheses. This issue will be discussed again in Phase 3, after students are introduced to the second type of problem they will study in this unit. So far, students have focused on only one of the two types of problems to which they will apply the χ^2 statistic—the case in which the null hypothesis is that a population fits a theoretical model. They will soon begin learning how to apply the χ^2 statistic to the case in which the null hypothesis is that two populations are the same.

Note: This is an important detail: For each of their two populations, students' survey of a sample should be equivalent to a "yes or no" question. For example, if students want to compare ninth graders and twelfth graders with regard to the type of music they like, they cannot use the χ^2 statistic to study whether the two groups share the same favorite type (in other words, they can't ask, "Is your favorite type of music rock, pop, rap, country, classical, or jazz?"). They can use it to study whether ninth graders and twelfth graders consider rock their favorite music type (in other words, they can ask, "Is rock your favorite type of music?"). This relates to the fact that students are studying the χ^2 statistic with *one degree of freedom.*

You may also want to emphasize that students' grades will be based partly on their meeting the deadlines in this activity. Unless their projects are unusual and they have prior approval, they need to have their plans in on schedule and their data collected in time for the in-class work period in Phase 6.

If time permits, you might have students begin the POW in class.

Phase 2: Choose Partners

Students turn in their partner selections. There may need to be one or more groups of three.

Phase 3: Choose a Hypothesis

Spend some time discussing the structure of hypotheses for the project. Students need to keep in mind two key issues:

While much of this unit deals with comparisons to a theoretical model, the POW asks for a hypothesis about *two* populations.

The survey needs to be a question with only two possible responses, equivalent to a "yes or no" question. If you completed *Big and Strong* before this phase of the POW, the examples there, along with the other 2-by-2 tables students saw in *What Would You Expect?,* should help clarify this issue.

Remind students that tomorrow (or whatever date you have chosen for Phase 4) they need to turn in a hypothesis, a null hypothesis, and a data-collection plan.

Phase 4: Turn in POW Hypotheses

Each team turns in a hypothesis, a null hypothesis, and a description of how they will collect data. Caution students not to begin collecting data until you have reviewed their plans.

Phase 5: Begin Data Collection

Return the hypothesis and null hypothesis statements and the data-collection plans. Remind teams that they must have their data ready for the day scheduled for Phase 6, when they will have class time to prepare their reports.

Phase 6: Complete POW Reports in Class

Give students at least 45 minutes in class to complete work on their reports. Remind them that completing their work includes preparing for their presentations and making a poster of their results. As you circulate, help teams decide how to use the χ^2 statistic, and suggest ways to clarify their conclusions.

Either assign a specific day for each team to present, or inform students that all teams need to be prepared by the day you have chosen for Phase 7 and that you will select teams at random to make presentations.

Phase 7: POW Presentations

Have the first set of student teams make their presentations. Encourage audience members to ask questions about how the presenters reached their conclusions. Ask questions yourself as needed to help presenters clarify their thinking.

If presenters reject their null hypothesis, ask, What alternate hypothesis do you think holds true? They should realize that the data set doesn't prove any particular alternate hypothesis.

If presenters do not reject their null hypothesis, ask, How would you explain any differences you saw in your populations?

You may want to photograph or videotape the presentations.

Late in the Day

Intent
This activity will help to assess students' understanding of the chi-square statistic.

Mathematics
In analyzing this real-world situation, students must recognize the theoretical model that is inherently suggested, form a null hypothesis, calculate a χ^2 statistic, find its associated probability, and make a decision. The activity is a good springboard for a discussion of the difference between *correlation* and *causation*.

Progression
Students will take information regarding accidents occurring on the job and make a decision as to whether time of day makes a significant difference. The follow-up discussion will raise the distinction between correlation and causation.

Approximate Time
25 minutes for activity (at home or in class)

10 minutes for discussion

Classroom Organization
Individuals or groups, followed by whole-class discussion

Materials
χ^2 *Distribution Graph* blackline master (transparency)

Doing the Activity
This activity requires little or no introduction.

Discussing and Debriefing the Activity
There isn't a theoretical model for the actual number of accidents that should take place in a given time period. As suggested in Question 2, however, it makes sense to think about *the fraction of the accidents* that should occur in each time period.

If "there is no difference" among parts of a shift in terms of likelihood of accidents, we would expect one-fourth of the accidents to occur during the last two hours of an eight-hour shift. In other words, one way to state the null hypothesis is, "If an accident is going to occur, there is a 25% chance that it will happen during the last two hours of a shift."

The total number of accidents is 168, so we would expect 42 accidents during the last two hours. A table for the situation might look like this.

	Last two hours	First six hours	Total
Number of	Expected: 42 Observed: 57	Expected: 126 Observed: 111	168

The usual computation gives a χ^2 statistic of approximately 7.14. Ask the class, What can the manager conclude from this χ^2 statistic? Students should use the χ^2 probability table to determine that if the null hypothesis were true, there would be less than a 1% chance of getting a χ^2 statistic this large or larger. Have students also look at where this χ^2 statistic fits on the χ^2 distribution curve, which will offer them a visual appreciation of how unusual the result is.

Students should conclude that it's very unlikely to be a coincidence that there are more accidents late in a shift. They might recommend that the manager adjust work schedules in some way, even though it may be expensive or difficult to do so.

This is a good time to clarify that students should do this type of analysis in their write-up of *POW 9: A Difference Investigation.* They will need to not only calculate the χ^2 statistic but also interpret it, using both the χ^2 probability table and a graph of the χ^2 distribution.

This is also a good occasion to raise the distinction between *correlation* and *causation.* For instance, you might ask, What specifically should the manager do about the fact that there seem to be more accidents late in a shift? Students may have some suggestions, but bring out that the correlation between "time in shift" and "number of accidents" doesn't necessarily point to a particular solution. It might be, for example, that workers start thinking about going home near the end of their shifts and pay less attention to safety. If that is the cause of accidents, it might not help to shorten the shifts. In other words, knowing that more accidents occur late in a shift does not explain what *causes* those accidents.

Key Questions

What can the manager conclude from this χ^2 statistic?

Where does each result fit in the χ^2 distribution?

What specifically should the manager do about the fact that there seem to be more accidents late in a shift?

Comparing Populations

Intent

Students' work with the chi-square statistic up to now has been exclusively in the context of comparison with theoretical probabilities. *Comparing Populations* extends their work to the two-population case.

Mathematics

One of the most powerful aspects of the chi-square statistic is that it applies to various categories of situations. In particular, it can be used both to compare observed data with theoretical probabilities and to compare data about two distinct populations. Applying the chi-square statistic to the two-population case is more complex.

In these activities, students will develop deeper intuitions about the meaning of the χ^2 statistic. They will use a simulation to estimate the χ^2 distribution, interpreting the χ^2 distribution curve as a probability table. They will calculate and interpret the χ^2 statistic to compare data from a real-world situation to a theoretical model and to compare two populations. And they will use the χ^2 statistic to make decisions, understanding that there are limitations to its application.

Progression

In *What Would You Expect?* students examine what the data would look like if two populations were the same with regard to a particular characteristic and if the null hypothesis were true. This work involves ideas about proportionality and can be quite challenging. Students work with this aspect of the chi-square statistic in several more situations, as well as make up their own data for given situations that would suggest rejecting the null hypothesis.

Students then learn how to adjust the chi-square statistic for the transition from the theoretical-probability case to the two-population case. In "*Two Different Differences" Revisited,* they apply their knowledge of the chi-square statistic to the two central unit problems. The remaining activities examine applications and misapplications of the statistic.

What Would You Expect?

Who's Absent?

Big and Strong

Delivering Results

Paper or Plastic?

Is It Really Worth It?

Two Different Differences Revisited

Reaction Time

Bad Research

On Tour with χ^2

What Would You Expect?

Intent

Students examine what would happen in situations comparing two populations if the null hypothesis were true.

Mathematics

Students encounter an important step in the use of the χ^2 statistic for the two-population case—finding the expected numbers. Doing so is more difficult in this case than in the theoretical-model case, so students will work on this step over several activities.

Progression

Students work in groups to find the expected numbers of right- and left-handed men and women in several groups of people, assuming gender makes no difference in handedness. The follow-up discussion focuses on ensuring that students understand the reasoning behind their calculations, specifically the roles of the null hypothesis and proportional reasoning.

Approximate Time

40 minutes

Classroom Organization

Groups, followed by whole-class discussion

Materials

Posters for "A Suspicious Coin" and "To Market, to Market" from *Two Different Differences*

Doing the Activity

Remind students that they've used the χ^2 statistic with only one of the *Two Different Differences* situations, "A Suspicious Coin," and ask, What's different about "To Market, to Market"? In that situation, they are comparing two populations to each other instead of comparing a population to a theoretical model.

Bring out the significance of this by asking, How did you use the theoretical model in "A Suspicious Coin"? Help them realize that it gave them the expected numbers, as what was expected was 50% heads and 50% tails.

In "To Market, to Market," students have two populations (men and women) and want to know if they truly differ *from each other* in some respect. Explain that finding expected numbers is somewhat more complicated in this case than in the problems they have been doing, and that the next several activities will focus on this aspect of the use of a null hypothesis and the χ^2 statistic. Students won't actually compute a χ^2 statistic for today's activity, because they will only be finding expected numbers, and they don't have any observed numbers.

Have students work on the activity in groups. If groups seem to be having trouble, you might bring the class together.

It may help to have students state the null hypothesis as precisely as possible. You might encourage them by asking, What fraction of the total sample is left-handed? If the null hypothesis were true, what fraction of the women would be left-handed? "Men and women are the same with respect to handedness" can also be stated as, "The fraction of men who are left-handed is the same as the fraction of women who are left-handed."

Discussing and Debriefing the Activity

During the discussion, it is important for students to recognize that proportions are involved and to understand how to calculate the numbers.

For Question 1, students can observe that 25% of the people are left-handed. Applying this fraction separately to men and women gives 25 right-handed women and 25 right-handed men. The completed table would look like this.

	Women	Men	Total
Right-handed	75	75	150
Left-handed	25	25	50
Total	100	100	200

Questions 2 and 3 are similar, although the arithmetic is a bit more difficult because the number of men does not equal the number of women.

	Women	Men	Total
Right-handed	90	60	150
Left-handed	30	20	50
Total	120	80	200

	Women	Men	Total
Right-handed	16.25	48.75	65
Left-handed	8.75	26.25	35
Total	25	75	100

Key Questions

What's different about "To Market, to Market"?

How did you use the theoretical model in "A Suspicious Coin"?

What is the null hypothesis?

What fraction of the total sample is left-handed? If the null hypothesis were true, what fraction of the women would be left-handed?

Who's Absent?

Intent

Students use proportional reasoning to find expected numbers in a two-population case.

Mathematics

This activity reinforces the mathematical ideas from *What Would You Expect?* In Question 3, students make up their own numbers, which will help you assess how well they understand the proportions involved.

Progression

Working on their own or in groups, students find the expected number of absences in a school for tenth graders and eleventh graders, given the total number of absences and the number of students in each grade.

Approximate Time

30 minutes for activity (at home or in class)

10 minutes for discussion

Classroom Organization

Individuals or groups, followed by whole-class discussion

Doing the Activity

Have a student read the introduction to the activity aloud. Make sure everyone understands which two populations are being compared and what attributes of those two populations are being compared.

As groups discuss Questions 1 and 2, circulate to determine whether it seems necessary to go over these answers.

Discussing and Debriefing the Activity

Here are the completed tables for Questions 1 and 2.

	Absent	Not absent	Total
10th graders	20	180	200
11th graders	20	180	200
Total	40	360	400

	Absent	Not absent	Total
10th graders	24	176	200
11th graders	36	264	300
Total	60	440	500

Ask students to report some of their examples for Question 3. This discussion should give you a sense of how well students understand the proportional reasoning involved.

If students are having trouble, you might suggest they set up the fractions in words. For example, for Question 1, if tenth graders and eleventh graders are alike with regard to absences, then these two fractions will be equal.

$$\frac{\text{total number of tenth graders}}{\text{total number of students}} \qquad \frac{\text{number of absent tenth graders}}{\text{number of absent students}}$$

Big and Strong

Intent

Students continue finding expected numbers based on the null hypothesis.

Mathematics

Many students have difficulty with the process of finding **expected numbers** in the two-population case. This activity offers more practice with the technique.

Progression

Working on their own, students find the expected numbers for two situations, based on the null hypothesis. Groups share their results through presentations. A follow-up discussion summarizes the process of finding expected numbers using proportional reasoning.

Approximate Time

25 minutes for activity (at home or in class)

10 minutes for discussion

Classroom Organization

Individuals or groups, followed by whole-class discussion

Doing the Activity

When introducing the activity, you might emphasize that without the assumption that the null hypothesis is true, there are many ways to complete these tables. However, only one solution in each situation fits this assumption exactly.

Discussing and Debriefing the Activity

Have students share their work in groups, and choose groups to present their results for each question. It is important that students understand how to calculate the expected numbers from the totals, so review the arithmetic as needed. Here are the completed tables (to the nearest tenth).

	Underweight babies	Babies not underweight	Total
Dr. Bertram	9.1	65.9	75
Other doctors	33.9	246.1	280
Total	43	312	355

	Good athletes	Not good athletes	Total
Dr. Pine	56.0	65.0	121
Other doctors	161.0	187.0	348
Total	217	252	469

Delivering Results

Intent

Students are introduced to computation of the χ^2 statistic for the two-population situation.

Mathematics

Students make an intuitive comparison of observed and expected numbers in the two-population case by choosing numbers that would support a particular hypothesis. This lays the foundation for introducing the use of the χ^2 statistic with two populations.

Progression

In the activity *Big and Strong,* students found the expected numbers in each of two situations, under the null hypothesis. Now, working on their own, they make up results for each situation that would support the doctors' claims. In a follow-up discussion, the teacher uses one set of students' proposed observed data to outline the use of χ^2 with two populations.

Approximate Time

25 minutes for activity (at home or in class)

20 minutes for discussion

Classroom Organization

Individuals, followed by whole-class discussion

Materials

χ^2 Distribution Graph blackline master (transparency)

Doing the Activity

When introducing the activity, ask students how this problem is similar to and different from previous two-population problems. Make sure they understand that they must select data that support the given outcomes.

If you are assigning this as homework, remind students that they will need their results from *Big and Strong.*

Discussing and Debriefing the Activity

Have one or two students propose observed numbers for each situation that they think would confirm the given hypothesis. Then ask, Why do you think your numbers prove the doctor's point? Point out that these made-up numbers have to go in the "right" direction, whereas in earlier problems, students simply found data they thought contradicted the null hypothesis.

Save one set of observed numbers for use later in discussing how to use the χ^2 statistic for the two-population case.

Correlation and Causation

This is another good opportunity to discuss the distinction between correlation and causation. The doctors claim that something about their methods of patient care leads to higher birth weight or increased athletic ability. You might spend a few minutes letting students speculate about what factors might be involved in these situations besides quality of medical care.

Computing χ^2 for Two Populations

Tell students they are now ready to apply the χ^2 statistic to decide whether two populations really differ in some respect based on differences in samples from the populations.

Ask students to outline the general procedure they have used in theoretical-model problems. What is a general outline of how to find and use the χ^2 statistic? They should come up with an outline something like this, which you may want to post.

State the null hypothesis.

Figure out what the data set would look like if it fit the null hypothesis. That is, calculate the expected values.

Compute the χ^2 statistic, comparing the observed and expected values.

Use the χ^2 probability table to find the probability associated with this χ^2 statistic.

Decide whether to reject the null hypothesis based on that probability.

Point out that in the last several problems (What Would You Expect?, Who's Absent?, and Big and Strong), students have been working on Step 2. They should recognize how this step is different in the two-population case from the theoretical-model case. Explain that the only other change between the theoretical-model case and the two-population case is a slight modification in how the χ^2 statistic is computed.

Use the student data saved earlier to demonstrate the calculation of the χ^2 statistic. The following discussion uses data for Question 1. The table shows how the data might look.

	Underweight babies	Babies not underweight	Total
Dr. Bertram	5	70	75
Other doctors	38	242	280
Total	43	312	355

Demonstrate this format for combining observed numbers and expected numbers, which students found in the activity Big and Strong.

	Underweight babies	Babies not underweight	Total
Dr. Bertram	Expected: 9.1 Observed: 5	Expected: 65.9 Observed: 70	75
Other doctors	Expected: 33.9 Observed: 38	Expected: 246.1 Observed: 242	280
Total	43	312	355

To calculate the χ^2 statistic, students will probably have a sense that they should continue to work with expressions of the form $\dfrac{(\text{observed} - \text{expected})^2}{\text{expected}}$, but may not be clear about how to do this in the new situation. You may need to clarify that they now have four fractions to add instead of two. You may want to point out that the general formula is

$$\sum \frac{(\text{observed} - \text{expected})^2}{\text{expected}}$$

Ask students to calculate the χ^2 statistic for Question 1 using the charts they have created. Here is the computation for the data in the previous chart.

$$\chi^2 = \frac{(5-9.1)^2}{9.1} + \frac{(38-33.9)^2}{33.9} + \frac{(70-65.9)^2}{65.9} + \frac{(242-246.1)^2}{246.1}$$

This gives a value of approximately 2.67. Students' values will likely be different, as they will be using their own sets of observed numbers.

The χ^2 Distribution Curve

Remind students that they developed the χ^2 distribution curve by examining situations in which they compared a population to a theoretical model, using data from Different Flips and Random but Fair.

Explain that even though the two-population case uses a slightly different computation for χ^2, the distribution curve for χ^2 is the same in that case as for the theoretical-model case. Bring out that because the distribution of χ^2 statistics is the same, the probability of getting a given χ^2 result is also the same. Therefore students can apply the χ^2 probability table they already have to the two-population case as well.

Using the χ^2 Curve and Table

Ask, What conclusion might the researcher come to based on your χ^2 statistic? For instance, if a student's result is close to the one just computed, he or she might say the researcher should reject the null hypothesis, reasoning that it's unlikely to be simply a sampling fluctuation that a smaller portion of Dr. Bertram's babies are underweight. (The probability associated with $\chi^2 = 2.67$ is approximately .10.)

As with previous problems, have students explain how they use their particular χ^2 statistic in their reasoning. They should refer to the placement of the result on the χ^2 distribution curve, as well as to the probability table that summarizes that curve.

Key Questions

Why do you think your numbers prove the doctor's point?

What is a general outline of how to find and use the χ^2 statistic?

What conclusion might the researcher come to based on your χ^2 statistic?

Supplemental Activity

Bigger Tables (extension) extends the ideas of expected numbers to larger tables than those in this activity. Students are expected to answer Question 2 based on their intuition. They revisit these data in the supplemental activity *A 2 Is a 2 Is a 2, or Is It?* in which they are introduced to χ^2 probability tables for more than one degree of freedom.

Paper or Plastic?

Intent

Students apply the chi-square statistic to a two-population problem.

Mathematics

This is the first of several activities in which students are required to synthesize all the steps of a χ^2 problem in the two-population case: extracting the observed data from a description of the situation, formulating hypotheses, organizing the data and calculating the expected results, computing the χ^2 statistic, and making a decision based on the probability associated with that statistic.

Progression

As this is the first time students are asked to put together all the steps in a two-population problem on their own, they work on the activity in small groups. The follow-up discussion reviews the use of the null hypothesis and the analogy to the presumption of innocence in the judicial system.

Approximate Time

30 minutes

Classroom Organization

Groups, followed by whole-class discussion

Doing the Activity

As groups work, circulate to make sure each group can figure out how to calculate the expected numbers and the χ^2 statistic.

Discussing and Debriefing the Activity

Have one or two students describe their reasoning for this activity. They probably will decide there is insufficient evidence to reject the null hypothesis, that is, there is insufficient evidence to conclude that there really is a difference in bag preferences. (The χ^2 statistic for this problem is about 1.07. There would be about a .30 probability of getting a χ^2 statistic this large or larger if the null hypothesis were true.)

Revisiting the analogy with the criminal justice system may help students understand. A juror will often say after a trial, "I voted not guilty even though I think the defendant is probably guilty. There simply wasn't enough evidence to convict." The standard for finding a defendant guilty of a crime is evidence that persuades the jury "beyond a reasonable doubt." If a juror thinks there is a 30% chance someone is innocent, there is definitely a "reasonable doubt," and the juror should vote to acquit, even though he or she believes the defendant is probably guilty.

Tell students this is how they should reason about getting results with this χ^2 statistic in the *Paper or Plastic?* situation. If the null hypothesis is true, the results seem slightly surprising, but we would still expect to get results this weird or weirder about one out of every three times we collect such data.

Now that students have used the χ^2 statistic in both the theoretical-model case and the two-population case, it may be worthwhile to summarize when to use the χ^2 probability table. Essentially, students can use this table as long as no expected value is less than 5 (the "χ^2 guideline") and they are working with one of two situations.

A single population is being compared with a theoretical model, and the characteristic being studied has two possible outcomes.

Two populations are being compared to each other, and the characteristic being studied has two possible outcomes.

Is It Really Worth It?

Intent

Students apply the chi-square statistic to a two-population problem.

Mathematics

This is the second of several problems that require students to follow the steps in conducting a χ^2 analysis in a two-population problem. There are no simple answers to the questions posed in this activity, illustrating that statistics do not eliminate the need for human judgment.

Progression

Students work on their own to apply the χ^2 statistic to a new situation, a process they should now be becoming comfortable with. The discussion emphasizes that despite the statistical calculation, human judgment is still needed in decision making.

Approximate Time

25 minutes for activity (at home or in class)

10 minutes for discussion

Classroom Organization

Individuals, then groups, followed by whole-class discussion

Doing the Activity

Read through the activity as a class. You may need to point out that the numbers in the table are the *observed* numbers, not the *expected* numbers.

Discussing and Debriefing the Activity

Have students work in groups to come to a consensus on the correct χ^2 statistic while you circulate to check that they are all able to do the χ^2 test correctly. As needed, have a student present the computation to the class. (The χ^2 statistic is approximately 2.6; the associated probability is about .11.)

Then ask groups to discuss what the veterinarian should do. It may be difficult for them to reach agreement. Spend a little time sampling group opinions and reasoning on this question, and bring out the observation that statistics do not eliminate the need for human judgment.

Supplemental Activity

TV Time (reinforcement) involves ideas similar to those in this activity and will give students more experience with finding expected numbers and using the χ^2 statistic in the two-population case.

Two Different Differences Revisited

Intent
Students make decisions by applying the χ^2 statistic, distinguishing as necessary between a two-population problem and a theoretical-model problem.

Mathematics
Students return to and apply their knowledge about the chi-square statistic to the two central problems of the unit. This requires that they distinguish between the two-population problem and the theoretical-model problem and recognize how to organize data and calculate χ^2 for each type of problem.

Progression
Students work in groups to return to the two situations in *Two Different Differences,* "A Suspicious Coin" and "To Market, to Market." The follow-up discussion will ensure that students understand and can distinguish between the two-population and theoretical-model problems.

Approximate Time
50 minutes

Classroom Organization
Groups, followed by whole-class discussion

Doing the Activity
Introduce *"Two Different Differences" Revisited* by explaining that it presents the same situations students examined earlier in this unit. They will now look at the situations using what they have learned over the course of the unit.

Students have been working recently with problems like "To Market, to Market," in which they compare two populations. You may wish to remind them that "A Suspicious Coin" is different in that it involves comparing a population to a theoretical model. Students already have computed the χ^2 statistic for "A Suspicious Coin" (refer to *Measuring Weirdness with χ^2*), but that was before they had the χ^2 probability table.

Mention that the write-up of this activity will be a required part of each student's portfolio.

Discussing and Debriefing the Activity
Discuss the conclusions students reached for each situation.

The χ^2 statistic for Roberto's brother's coin is about 21.3. The first time students found this number, they did not have the χ^2 probability table. Now they can recognize that this value is off the table, so they know that the probability of such lopsided results with a fair coin is less than .005.

For "To Market, to Market," students will likely have observed that 60% of the men preferred the new soft drink while only 55% of the women did. A completed table of observed and expected values will look something like this.

	Prefer new soft drink	Prefer old soft drink	Total
Men	Expected: 52.2 Observed: 54	Expected: 37.8 Observed: 36	90
Women	Expected: 34.8 Observed: 33	Expected: 25.2 Observed: 27	60
Total	87	63	150

The χ^2 statistic comes out to about 0.37, with an associated probability of about .565. Therefore, it is very likely that the difference observed between men and women is due to a sampling fluctuation, and there is insufficient evidence to reject the null hypothesis.

Supplemental Activities

Degrees of Freedom (extension) addresses the concept of *degrees of freedom,* which is needed to extend the use of the χ^2 statistic to tables larger than 2 by 2.

A 2 Is a 2 Is a 2, or Is It? (extension) continues work with the concept of degrees of freedom and can be assigned after students have completed the supplemental activity *Degrees of Freedom.*

Reaction Time

Intent

This activity and *On Tour with* χ^2 provide opportunities for students to pull the main ideas of the unit together.

Mathematics

Students analyze data and come to conclusions using the χ^2 statistic in another two-population problem.

Progression

Working on their own, students follow the stages of statistical investigation for a new situation. They make a hypothesis, state the null hypothesis, calculate the chi-square statistic, and then find its associated probability from the chi-square probability table. Students share findings in a class discussion.

Approximate Time

25 minutes for activity (at home or in class)

10 minutes for discussion

Classroom Organization

Individuals, followed by whole-class discussion

Doing the Activity

This activity requires little or no introduction.

Discussing and Debriefing the Activity

By now, this type of problem should be fairly routine. However, be sure to have one or two students present their work to the class so that those who are still uncomfortable with some aspects of the process can ask questions.

Bad Research

Intent

Students critique a situation involving a misapplication of the chi-square statistic.

Mathematics

This activity approaches silliness, as the mistakes depicted are rather extreme, but working on it will help students synthesize some of the principles of methodology they have learned.

Progression

Presented with conclusions based on poor research, students work in groups to identify and critique the faulty research techniques, expressing and reinforcing their understanding of what constitutes good research. In the follow-up discussion, students share what is wrong with the research described in the activity.

Approximate Time

30 minutes

Classroom Organization

Groups, followed by whole-class discussion

Doing the Activity

Have students work on the activity in groups. You might have group members pool their answers to find out which group can come up with the longest list of mistakes the researcher made.

Discussing and Debriefing the Activity

You might ask the group that claims to have found the most mistakes to read its list to the class and then allow other groups to suggest additional ideas.

Here are some mistakes students might point out.

One should not thumb through an associate's files without permission.

The researcher assumes there are equal numbers of boys and girls. In fact, there is no evidence there are *any* girls at the school.

The person data-snooped and then made a hypothesis, but did not check it with a new sample.

The researcher does not state a null hypothesis.

The statistic is "chi-square," not "ex-square."

Chi-square should not be used when any expected value is less than 5.

In computing the χ^2 statistic, the researcher divides by the observed number instead of the expected number.

Division by zero is undefined, not zero.

The 22% represents how often the χ^2 statistic would be 1.52 *or larger* if the null hypothesis were true, not the probability of getting that particular value.

Using this 22% to describe how many more boys than girls fail is nonsense.

The researcher generalizes to populations not represented in the sample.

The researcher erroneously applies cause and effect to not being smart and failing classes.

There will probably be some debate as to whether these are all really mistakes.

On Tour with χ^2

Intent

This activity gives students another problem where they can pull the main ideas of the unit together.

Mathematics

This is a wonderful final opportunity to ensure that students are comfortable with all the stages in a χ^2 investigation prior to doing the analysis for their own research in *A Difference Investigation.*

Progression

Use this final activity, which students will work on individually, to assess whether any aspect of the χ^2 investigation is still giving students difficulty.

Approximate Time

25 minutes for activity (at home or in class)

10 minutes for discussion

Classroom Organization

Individuals, followed by whole-class discussion

Doing the Activity

This activity requires little or no introduction.

Discussing and Debriefing the Activity

Have presentations and discussion as needed. You may want to focus the discussion on Questions 1d and 2d, which ask how students would recommend improving the studies.

Question 1 gives a χ^2 statistic of about 3.83, which has an associated probability of about .05. Question 2 gives a χ^2 statistic of about 4.67, which has an associated probability of about .03.

POW Studies

Intent

These activities are devoted to final preparation and presentations of students' work on *POW 9: A Difference Investigation.*

Mathematics

The main activity in *POW Studies* is for students to complete work on their reports for *A Difference Investigation* and to assemble their portfolios. The POW is an authentic assessment of each student's ability to follow the stages of investigation from beginning to end.

Progression

Wrapping It Up

Beginning Portfolio Selection

Is There Really a Difference? Portfolio

Wrapping It Up

Intent

Students complete work on their reports for *POW 9: A Difference Investigation.*

Mathematics

Students will use the χ^2 statistic to make decisions, applying the stages of statistical investigation learned in the unit. They will complete a long-term project based on a research question of their choice, demonstrating good sampling and statistical techniques.

Progression

Students worked in class to complete their analyses for their POWs. This activity is simply a reminder for them to complete their written reports as homework.

Approximate Time

1 hour (at home)

Classroom Organization

Individuals

Doing the Activity

If students have finished their work with their partners in class and are ready to present their findings, they may not have any homework tonight.

Assign specific days for presentations, or tell all teams to be ready tomorrow.

Beginning Portfolio Selection

Intent
Students begin compiling their unit portfolios.

Mathematics
This activity is an opportunity for students to reflect about the types of decisions for which the χ^2 statistic might be useful.

Progression
Students are asked to review their work from this unit and to think about when the χ^2 statistic is or is not useful. This activity will be included in their portfolios.

Approximate Time
40 minutes for activity (at home)

10 minutes for discussion

Classroom Organization
Individuals, followed by whole-class discussion

Discussing and Debriefing the Activity
Ask a few students to share their ideas with the class. You may want to make a chart with two columns—one for situations in which students feel the χ^2 statistic is useful and one for situations in which they feel it is not useful.

Is There Really a Difference? Portfolio

Intent

Students complete their unit portfolios and write their cover letters.

Mathematics

Students' portfolio selections will include the artifacts they identified in *Beginning Portfolio Selection.*

Progression

Students start work on their portfolios in class by reading the instructions in the student book. They then work independently to write a summary of the mathematical ideas from the unit, gather some specified completed activities and select others, and reflect upon how *POW 9: A Difference Investigation* helped them to understand and to appreciate the key concepts of the unit.

Approximate Time

5 minutes for introduction

45 minutes for activity (at home)

Classroom Organization

Individuals

Doing the Activity

Have students read the instructions in the student book carefully. They have done part of the selection process in *Beginning Portfolio Selection,* so their main task now is to write their cover letters.

Discussing and Debriefing the Activity

You may want to have a few students read their cover letters summarizing the key ideas of the unit. It might also be productive to compile a class list of ideas and terminology that students learned and worked with in this unit. Here are some items that might be on such a list.

Frequency bar graph

Double-bar graph

Sample

Population

Data snooping

Hypothesis making

Hypothesis testing

Theoretical model

Null hypothesis

"Rejecting the null hypothesis"

Sampling fluctuation

χ^2 statistic

Expected numbers

Observed numbers

Measuring "weirdness"

Probability

Standard deviation

Normal distribution

Blackline Masters

Loaded Dice

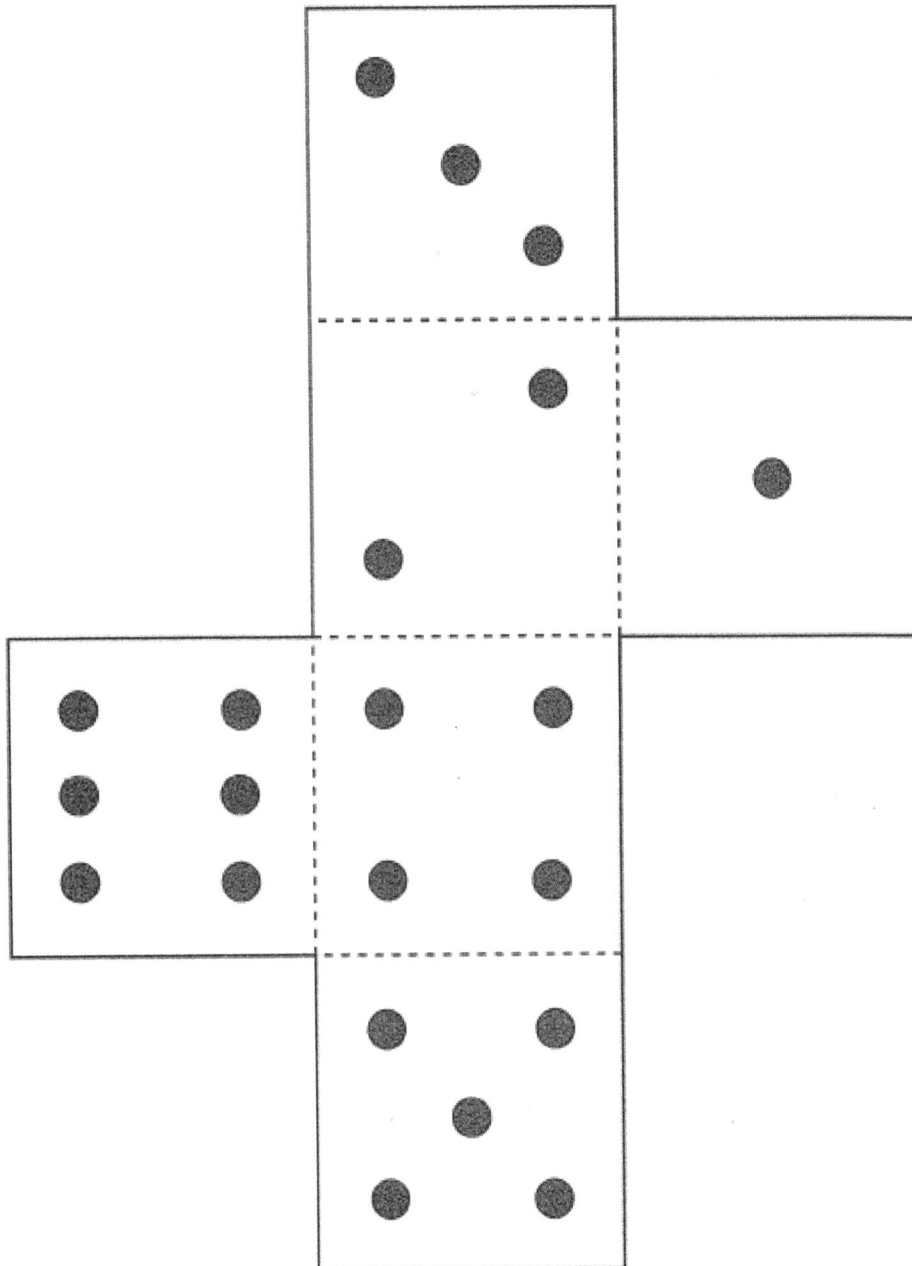

χ^2 Distribution Graph

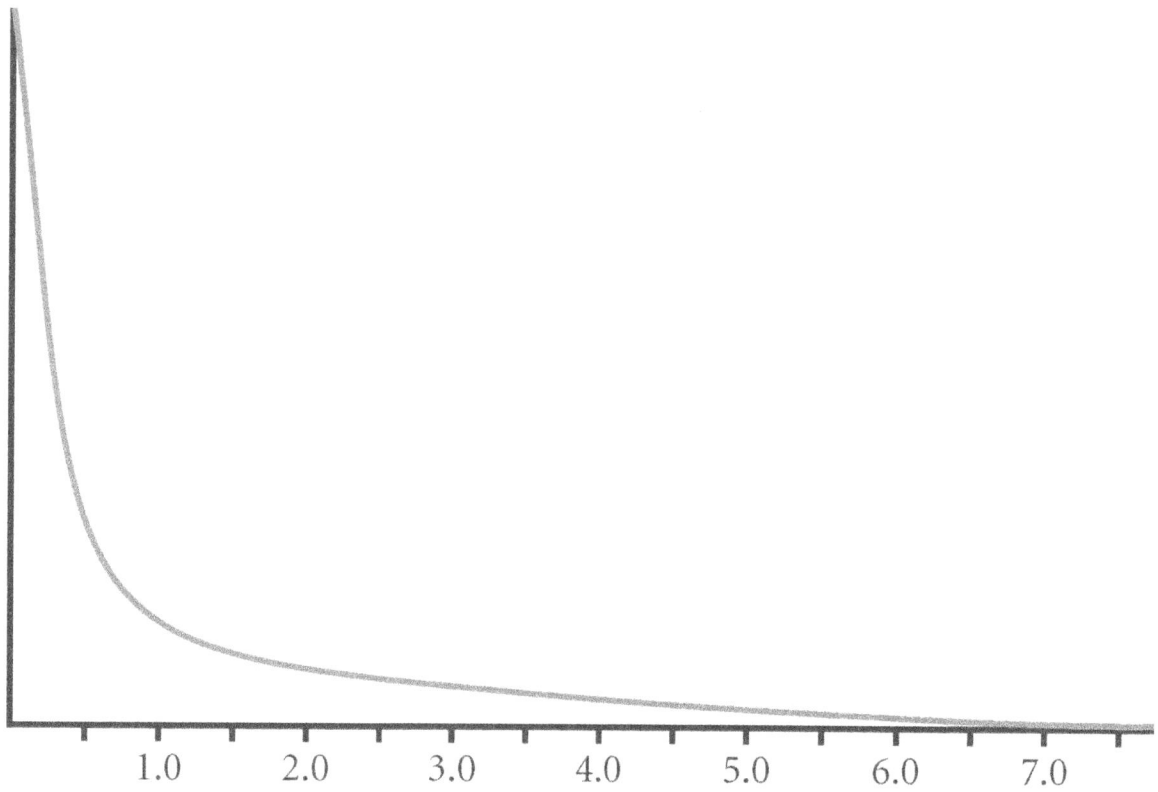

Schedule for POW 9: A Difference Investigation

The stages of your work on the POW *A Difference Investigation* will follow this schedule.

On _____, you tell your teacher who your partner is.

On _____, you and your partner hand in

- a precise statement of your hypothesis
- a statement of your null hypothesis
- a method for collecting sample data from each of your populations. If your method of data collection involves a questionnaire, include a copy of the questionnaire.

Between _____ and _____, you and your partner collect your data.

On _____, you hand in your raw data.

On _____, you and your partner analyze your results and prepare a poster and a presentation.

On _____, you hand in your own write-up of the POW.

On _____, you and your partner make your presentation using your poster.

Assessments

In-Class Assessment

An auto dealer is listening one morning to a radio talk show about car sales. A caller claims that women select cars primarily based on appearance and that men select cars primarily based on performance.

The auto dealer wants to know whether there is any validity to the caller's claim. So the dealer conducts a survey, which asks people this question:

What one characteristic of your present car is most important to you?

A total of 200 men and 250 women respond to the survey. Of those, 81 men and 126 women give an appearance characteristic. The rest give a performance characteristic.

1. State a null hypothesis for this situation. Define the appropriate populations clearly.
2. Precisely state the caller's claim in terms of the populations described in Question 1.
3. Make and label a chart that will help you calculate the χ^2 statistic for this survey.
4. Compute the χ^2 statistic.
5. Do you think the results of this survey support the caller's claim? Explain your answer carefully, using the χ^2 probability table.
6. What flaws might this study have? How could they be corrected?

Take-Home Assessment

Ming is a champion surfer. She is often asked by companies to say their products are the best.

Ming was most recently asked by several surfboard-wax companies to endorse their brands of wax. Surfers put wax on their surfboards to keep from slipping. The wax makes the surfboards feel sticky.

Ming is unwilling to say that a certain wax is best unless she is quite certain it is. So she decides to collect some data. Being statistically sophisticated, she begins with a hypothesis and a null hypothesis.

She hypothesizes that users of Footglue wax have fewer slips than users of other brands of surfboard wax. She then tests her hypothesis at a surfing competition. She found out in advance which surfers were using Footglue wax and which were using other brands.

Of a total of 200 rides made by surfers using Footglue wax, 9 involved slips and 191 did not. For users of other brands, out of 600 rides, 36 involved slips and 564 did not.

Should Ming say that Footglue wax is best?

Your report should include a null hypothesis, a χ^2 analysis, and an explanation of what the χ^2 statistic means in the context of this problem.

Is There Really a Difference? Guide for the TI-83/84 Family of Calculators

This guide gives suggestions for selected activities of the Year 2 unit *Is There Really a Difference?*. The notes that you download contain specific calculator instructions that you might copy for your students. NOTE: If your students have the TI-Nspire handheld, they can attach the TI-84 Plus Keypad (from Texas Instruments) and use the calculator notes for the TI-83/84.

For most of this unit, students will be using little more than the basic calculator functions. They will review standard deviation calculations with the calculator, but they will not devote much time to them. The activities *Coin-Flip Graph* and *Graphing the Difference* provide opportunities to explore constructing frequency bar graphs on the calculator, either as a class if there is time or on an individual basis for advanced students if there is not. In *Random but Fair,* students will learn to use the random number generator to simulate a probability situation.

The chi-square calculation itself will provide some reinforcement of order-of-operations rules on the calculator. Although students need to become familiar with the step-by-step mechanics of calculating chi-square statistics, we have a few suggestions to make life easier for you. For those of you who have advanced students working with problems beyond a single degree of freedom, we discuss the calculator's built-in chi-square test. We also provide a program for teachers to use that will greatly reduce the amount of time needed to check the accuracy of students' work in *POW 9: A Difference Investigation.*

The Dunking Principle: The probabilities from this activity can be illustrated vividly on the overhead projector calculator. Use the **randBin** function, which is found in the MATH **PRB** menu. To simulate pushing the button at the dunking booth 20 times, with the button set to dunk the victim 50 percent of the time, enter **randBin(20,.5)**. Each time you press ENTER, the calculator will display a new number of dunks out of 20 trials. You can quickly illustrate the improbability of getting 15 or more dunks out of 20 or getting 46 or more out of 60.

Normal Distribution and Standard Deviation: Though students calculated standard deviation using the calculator in *The Pit and the Pendulum* in Year 1, they will probably need a review of the details of doing so. The note "Finding the Standard Deviation with the Calculator" assumes that students have the coin flip data sets that they used to construct a frequency bar graph for the activity *Coin-Flip Graph.* This calculator note tells students how to enter statistical data into the calculator and how to calculate standard deviation. If you wish to have students construct a chi-square distribution graph as an extension after the activity *Graphing the Difference,* tell them not to clear this data set from the lists in their calculators after that day's activity.

If you have extra time, you might introduce using the calculator to quickly construct a bar graph. The calculator note "Constructing a Bar Graph on the Calculator" explains how to use the calculator to create the graph from *Coin-Flip Graph.*

How Does χ^2 Work?: As students begin to calculate the chi-square (χ^2) statistic, you may need to remind them about the need to use parentheses to control the order of operations. A common mistake is neglecting to use parentheses when squaring a negative difference between the observed and expected values. Point out that the result of squaring the difference in the numerator will always be a positive number or zero. If the calculator seems to say differently, chances are the student forgot to use parentheses. If further reinforcement of this idea seems warranted, you might have students verify the equivalence of two expressions like $(X - 5)^2/10$ and $(5 - X)^2/10$ by graphing or examining the tables.

Graphing the Difference: You may want to ask a group that finishes work on this activity early to find out if they can duplicate their bar graph on the overhead projector calculator. The calculator note "Constructing a Bar Graph on the Calculator" will be helpful.

If you have time, and if you still have the data set from *Coin-Flip Graph* in the calculator, it may be interesting to construct the χ^2 distribution graph from that data set to compare with the data set from *Graphing the Difference.*

You can calculate the χ^2 values for the entire list of results (**L1**) at once by entering the following keystroke sequence on the home screen: `(` 2nd [L1] `–` 5 0 `)` x^2 `÷` 5 0 STO> 2nd [L2]. This calculates χ^2 for each experiment from List 1 and records χ^2 in List 2.

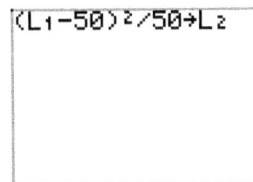

```
(L1–50)²/50→L2
```

Press 2ND [STAT PLOT] and select **Plot1**. Highlight **On** and the histogram symbol, and then set **Xlist** to **L2** and **Freq** to **1**.

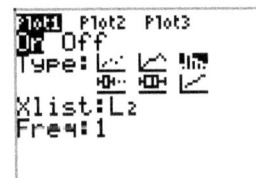

```
Plot1 Plot2 Plot3
On Off
Type:
Xlist:L2
Freq:1
```

Press WINDOW and set the values with **Xmin** equal to 0 and **Xmax** at about 9. Set **Xscl** to the desired interval for each bar, probably about 0.2. To make comparison more meaningful, you may want to use the same interval that you used for *Graphing the Difference.* (Remember the expression **(Xmax – Xmin)/Xscl** cannot exceed 47 due to screen width limitations.)

You will need to adjust **Ymin** and **Ymax** to appropriate values for the number of experiments in your data set.

Press GRAPH to view the histogram.

Random but Fair: In this activity, students use the calculator's random number generator to simulate the 70% probability of getting a good tip from the activity *Does Age Matter?* The procedure described in the calculator note "Generating a Random Seed Number" will ensure that random numbers are not determined by a seed number preset at the factory.

POW 9: A Difference Investigation: As you assist students with this POW and, ultimately, as you evaluate their final results, the program presented in the calculator note "Running the Chi-Square (χ^2) Program" will be very useful. Given the *observed* values for the four cells of the table, the program calculates the *expected* values for each cell and calculates the χ^2 statistic. The observed values can be entered in either row or column order. The program uses a setup like this for the data.

Cell 1	Cell 2	row sum
Cell 3	Cell 4	row sum
column sum	column sum	grand total

It is strongly suggested that you not make this program available to students. Although it can be a wonderful tool for teachers, students need to wrestle with the process of determining expected values and calculating χ^2 in order to understand the concepts fully. For that reason, although this program is presented on separate pages for your convenience, these pages are intended for teacher use only. You can also download this program and load into onto your calculator.

The calculator also has a built-in feature for calculating χ^2, but for problems with a single degree of freedom, that feature is less convenient to use than the program given here. However, you may find the feature useful if some students are pursuing more complex problems involving more than one degree of freedom. Instructions for using the built-in χ^2 test are provided in the calculator note "Using the χ^2 Test." These instructions are also intended for teacher use only.

Big and Strong: When working with intermediate answers on the calculator, students often fall into one of two bad habits: either they round all intermediate answers to integer values before reentering them into the calculator for the next calculation, or they try to reenter all ten digits displayed for the intermediate answer.

While reviewing the mechanics of calculating the expected numbers for *Big and Strong,* encourage students to leave intermediate answers in the calculator. For example, students may have begun the process of finding the expected number of underweight babies for Dr. Bertram by dividing 75 by 355 to discover what percentage of the babies Dr. Bertram delivered. If they round the result to 21 percent and then multiply this by the column total of 43 underweight babies, they will find 9.0 (to the nearest tenth) as the expected number of underweight babies for Dr. Bertram. But if they simply leave the result of dividing 75 by 355 in the calculator without rounding and multiply that by 43 (as shown here), they will get a more accurate answer, which rounds to 9.1.

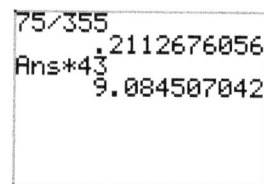

```
75/355
           .2112676056
Ans*43
           9.084507042
```

Calculator Notes

Finding the Standard Deviation with the Calculator

These instructions will show you how to enter the coin toss data set that you used in *Coin–Flip Graph,* and how to find the standard deviation of that data set using the calculator.

Entering Statistical Data Sets

Press STAT to bring up the menu for entering data. To clear any previous data sets from the calculator, simply press ENTER, since **1:Edit** is already selected. Use the up arrow key to move the cursor onto the **L1** heading at the top of the screen. Press CLEAR ENTER to clear list **L1**.

The data set will be entered into list **L1**. Move the cursor to the next line below the **L1** heading and enter each of the numbers of heads that were observed. After each entry, press ENTER.

Calculating Standard Deviation Step-by-Step

Your data set should already be entered in list **L1**, as described previously. Here are the steps for calculating the standard deviation of your data set.

1. Find the mean.

2. Find the difference between each data item and the mean.

3. Square each of the differences.

4. Find the average (mean) of these squared differences.

5. Take the square root of this average.

Doing these steps (especially repeating steps 2 and 3 for each individual data item) would be very time consuming. Fortunately, the calculator offers some shortcuts.

1. **Find the mean.** To find the mean of the data, press 2ND [**LIST**] and use the right arrow key to highlight the **MATH** menu heading. Highlight 3:**mean(** and press ENTER. To specify that list **L1** contains the data set whose mean you wish to find, press 2ND [**L1**]. Add the closing parenthesis. Finally, tell the calculator to store the result as *X* by pressing STO> X,T,θ,n ENTER. The calculator will display the mean.

```
NAMES OPS MATH
1:min(
2:max(
3▓mean(
4:median(
5:sum(
6:prod(
7↓stdDev(
```

```
mean(L₁)→X
        49.14285714
```

2. **Find the difference between each data item and the mean.** Pressing 2ND [L1] − X,T,θ,n STO> 2ND [L2] ENTER tells the calculator to subtract the mean from each data item in list **L1** and to store the result in list **L2**.

```
L₁-X→L₂
{-13.14285714 3...
```

3. **Square each of the differences.** Tell the calculator to square each data item in list **L2** and to store those squares in list **L3** by pressing 2ND [L2] x² STO> 2ND [L3] ENTER.

```
L₂²→L₃
{172.7346939 14...
```

4. **Find the mean of the squared differences.** Take the mean of list **L3** by selecting **3:mean** from the 2ND [LIST] MATH menu and specifying **L3**.

5. **Take the square root of this average.** Do this by pressing 2ND [√] 2ND [ANS] ENTER.

```
mean(L₃)
        35.55102041
√(Ans)
        5.962467644
```

Calculator Shortcut for Standard Deviation

The calculator provides a much quicker way of calculating standard deviation than by performing each of the individual steps.

1. Press STAT and use the arrow keys to move to the **CALC** menu.

2. Select **1-Var Stats** and press ENTER, so that the command appears on your home screen.

3. Press 2ND [L1] after the command to specify that you want to calculate statistics for the data set in list **L1**, and then press ENTER to display the statistics.

```
EDIT CALC TESTS
1▓1-Var Stats
2:2-Var Stats
3:Med-Med
4:LinReg(ax+b)
5:QuadReg
6:CubicReg
7↓QuartReg
```

```
1-Var Stats L₁
```

```
1-Var Stats
x̄=49.14285714
Σx=344
Σx²=17154
Sx=6.440201121
σx=5.962467644
↓n=7
```

A number of values will be displayed. Those of most interest at this time include these:

$\bar{x}$: the mean of the data set

σ**x**: the standard deviation

n: the number of items in the data set

Constructing a Bar Graph on the Calculator

These instructions will tell you how to use the calculator to re-create the frequency distribution graph that you drew as part of *Coin Flip Graph*.

1. If you have not yet entered the data set from that assignment, do so at this time.

2. Press $\boxed{Y=}$ and clear any functions that are entered there so that they will not interfere with the bar graph.

3. Access the setup screen for the graph by pressing $\boxed{2ND}$ [**STAT PLOT**]. If **Plot2** or **Plot3** are **On**, turn them off by selecting **PlotsOff** and then pressing $\boxed{ENTER}$ when you see **PlotsOff** on the home screen. Then press $\boxed{2ND}$ [**STAT PLOT**] again.

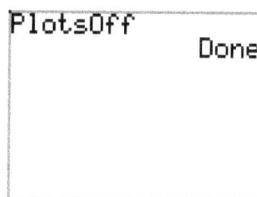

```
STAT PLOTS              PlotsOff
1:Plot1…On                        Done
   ⊾  L1   L2   ▫
2:Plot2…Off
   ⊾  L1   L2   ▫
3:Plot3…Off
   ⊾  L1   L2   ▫
4↓PlotsOff
```

4. Select **Plot1** from the Stat Plot screen by pressing $\boxed{ENTER}$. Use the arrow keys to highlight **On**, and press $\boxed{ENTER}$ to turn the plotting function on. Use the arrow and $\boxed{ENTER}$ keys similarly to select the bar graph symbol next to **Type**. Set **XList** to **L1** (using $\boxed{2ND}$ [**L1**]), and set **Freq** to **1** (press $\boxed{ALPHA}$ to exit the ALPHA mode and then press $\boxed{1}$).

```
Plot1  Plot2  Plot3
On  Off
Type: ⊾  ⊿  ᴧ
      ⊞  ⊞  ⊿
Xlist:L1
Freq:1
```

5. Press $\boxed{WINDOW}$ and set up the viewing window for your display. Set **Xmin** and **Xmax** to reflect the range of values for the number of heads observed in the 100-toss experiments. Set **Ymin** to 0 and **Ymax** to slightly more than your greatest frequency. **Xscl** will be the interval for each bar. Unless you want to group your data into intervals, use an interval of 1. (The expression (**Xmax – Xmin**)/**Xscl** cannot exceed 47 due to screen width limitations.) Press $\boxed{GRAPH}$ to view your graph.

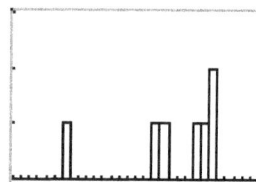

6. Press $\boxed{TRACE}$ to observe the values for each bar. The variables **min** and **max** define the range of values for each bar, and **n** is the frequency for that bar. Note that any data items that fall on the boundary between two bars will be plotted as part of the bar to the right.

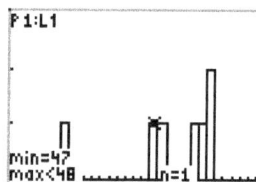

```
P1:L1



min=47
max<48        n=1
```

Generating a Random Seed Number

In *Random but Fair*, you are asked to use the calculator's random number generator to simulate the 70% probability of Clementina getting a good tip if her null hypothesis is true. But the random number generator on a calculator is not truly random. It needs a seed value from which to start. If you do not supply a seed value, the random number generator will start from a value preset at the factory, and as a result you will have the same set of "random" numbers as everyone else in the class. To avoid this, follow these steps.

1. Enter a handful of random digits and store them as the seed by pressing STO> MATH. Use the arrow keys to select the **PRB** (probability) menu and press ENTER to select **1:rand**. Press ENTER again to store this seed value.

```
548192625→
```

```
MATH NUM CPX PRB
1:rand
2:nPr
3:nCr
4:!
5:randInt(
6:randNorm(
7:randBin(
```

```
548192625→rand
         548192625
```

2. Generate the first random number by pressing MATH, again selecting **1:rand** in the **PRB** menu, and pressing ENTER.

```
548192625→rand
         548192625
rand
       .9318259094
       .8255144227
       .9776168475
       .9259235368
```

3. Generate the remaining random numbers by repeatedly pressing ENTER.

Running the Chi-Square (χ^2) Program

These instructions are intended for teacher use only. At this stage, students benefit from calculating expected values and χ^2 step by step.

Begin by pressing [PRGM] and selecting **Create New** under the **NEW** menu. Enter the program name, **CHISQUAR**, and press [ENTER]. More detailed instructions for entering programs can be found in the calculator note "Introduction to Programming the Calculator," which is provided with the Calculator Notes for the Year 2 unit *Do Bees Build it Best*.

```
PROGRAM
Name=CHISQUAR
```

Instruction	Explanation
:ClrHome	Clears the screen. Enter this command by pressing [PRGM] and then using the right arrow key to select the **I/O** (input/output) menu heading. You may need to scroll off the bottom of the screen to find **8:ClrHome**.
:Disp "CELL 1 OBSERVED"	Prompts the user to input the observed value for Cell 1. Like the previous command, **Disp** (display) is found on the **I/O** menu. Press [2ND] [**A-LOCK**] to enter the rest of the line. Enter the spaces using the ALPHA character above the zero key; enter the quotation marks using the ALPHA character above the [+] key.
:Input A	Inputs the observed value for Cell 1. Again, the **Input** command is selected from the **I/O** menu. Remember to use the [ALPHA] key to add the variable **A**.
:Disp "CELL 2 OBSERVED"	This and the next several commands display prompts to input observed values for Cells 2 through 4.
:Input B	
:Disp "CELL 3 OBSERVED"	
:Input C	
:Disp "CELL 4 OBSERVED"	
:Input D	

:A+B→E	Calculates the total of the observed values in Row 1. Enter → with the STO> key.
:C+D→F	Calculates the total of the observed values in Row 2.
:A+C→G	Column 1 total.
:B+D→H	Column 2 total.
:E+F→T	Total of observed values in all four cells.
:(E/T)*G→W	Calculates the expected value for Cell 1.
:E−W→X	Expected value for Cell 2.
:G−W→Y	Expected value for Cell 3.
:F−Y→Z	Expected value for Cell 4.
:(A−W)^2/W+(B−X)^2/X+(C−Y)^2/Y+(D−Z)^2/Z→S	
	Calculates χ^2.
:ClrHome	Clears the screen again. Press PRGM and select this from the **I/O** menu.
:Output(1,1,"EXPECTED VALUES")	
	The next several commands label and display the expected values and χ^2. The **Output** command controls the placement of the information on the calculator screen. Press PRGM and select this from the **I/O** menu.
:Output(2,1,W)	
:Output(3,1,X)	
:Output(4,1,Y)	
:Output(5,1,Z)	
:Output(6,1,"CHI SQUARE")	
:Output(7,1,S)	

:Stop

Select **Stop** from the **CTL** (control) menu after pressing PRGM. You may have to scroll below the bottom of the screen to find it.

Complete program entry by pressing 2ND [QUIT].

To run the program, press PRGM, select the name of the program in the **EXEC** (execute) menu, and press ENTER twice. The first screen display shows a calculator user responding to the program by entering numbers at each question mark. The second screen display shows what will happen when you press ENTER after entering the number 30.

```
CELL 1 OBSERVED
?15
CELL 2 OBSERVED
?17
CELL 3 OBSERVED
?28
CELL 4 OBSERVED
?30
```

```
EXPECTED VALUES
15.28888889
16.71111111
27.71111111
30.28888889
CHI SQUARE
.0162197785
```

Using the χ^2 Test

As with the previous program, these instructions are intended for teacher use only. At this stage, students benefit from calculating expected values and χ^2 step by step.

Some of your students may raise interesting questions that involve χ^2 calculations with more than a single degree of freedom. This note will help you use the calculator's χ^2 test feature to check such student work.

Enter the observed values into matrix A. Press 2ND [**MATRIX**], select the **EDIT** menu heading, and press 1 or ENTER to select matrix A.

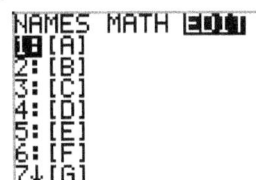

At the top of the screen, enter the matrix dimensions as the number of rows followed by the number of columns. Press ENTER after each entry to move to the next. Enter the observed value for each cell into the matrix.

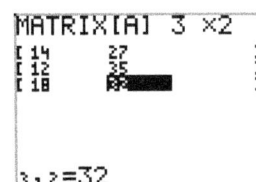

To calculate χ^2, press STAT, select the **TESTS** menu heading, and scroll down to **C**: χ^2-**Test**. Press ENTER. The default selections will appear: matrix **[A]** for the observed values and matrix **[B]** for the expected values. If you wish to use different matrices, move the cursor to the matrix to be replaced, press 2ND [**MATRIX**], and select the desired matrix from the **NAMES** menu.

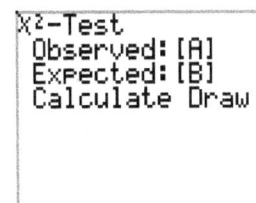

Highlight **Calculate** and press ENTER. The calculator will display the resulting χ^2 value, as well as the probability of obtaining data that shows this much of a difference from the expected values if the null hypothesis is true (that is, if there is no association between row variables and column variables). The number of degrees of freedom is represented by **df**.

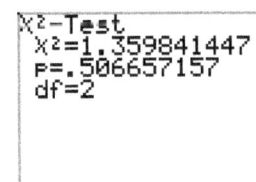

If you wish to look at the expected values, press 2ND [**MATRIX**] and select **[B]** from the **NAMES** menu. When **[B]** appears on the home screen, press ENTER. The expected values in matrix **[B]** will be displayed. The dots at the right-hand edge of the screen mean that you can use the right arrow key to scroll the screen in that direction to observe the cells that do not fit on the screen.

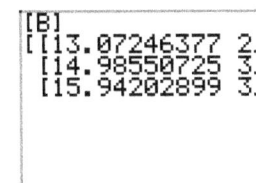

The calculator can also display a χ^2 distribution graph with a shaded region that contains χ^2 values this large or larger. Press STAT, highlight the **TESTS** menu heading, and select

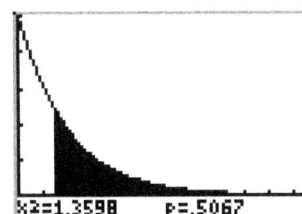

χ^2-**Test** again. This time, select **Draw** instead of **Calculate**. (You probably want to prevent other graphs from interfering with the χ^2 distribution graph. To do that, press Y= and turn off the graphs of any functions entered there; also, press 2ND [**STAT PLOT**], select **PlotsOff,** and then press ENTER.)

www.ingramcontent.com/pod-product-compliance
Lightning Source LLC
Chambersburg PA
CBHW051346200326
41521CB00014B/2495